Understanding and Training Your Cat or Kitten

Understanding and Training Your Cat or Kitten

H. Ellen Whiteley, D.V.M.

Illustrations by Jennifer Harper

Sunstone books may be purchased for educational, business, or sales promotional use. For information please write: Special Markets Department, Sunstone Press, P.O. Box 2321, Santa Fe, New Mexico 87504-2321.

Library of Congress Cataloging-in-Publication Data:

Whiteley, H. Ellen.
Understanding and training your cat or kitten / H. Ellen Whiteley.
p. cm.
Includes index.
Originally published: New York : Crown Trade Paperbacks, c1994.
ISBN: 0-86534-509-0 (alk. paper)
1. Cats--Behavior. 2. Cats--Training. I. Title.

SF446.5.W48 2006
636.8--dc22

2006042399

WWW.SUNSTONEPRESS.COM
SUNSTONE PRESS / POST OFFICE BOX 2321 / SANTA FE, NM 87504-2321 /USA
(505) 988-4418 / ORDERS ONLY (800) 243-5644 / FAX (505) 988-1025

To my patients, friends, clients, and colleagues in the writing and veterinary professions. Thank you for sharing your knowledge and understanding with me. And to C.K., in loving memory.

Contents

Preface
1

1
People and Cats – 5

2
Starting Right – Kitten or New Cat – 19

3
Sensing, Learning, and Communicating – 33

4
Etiquette Training – 51

5
Country Cat, City Cat – 69

6
The Basics – 95

7
Teaching – 113

8
Correcting Misbehavior — 129
9
Stress — 155
10
Diseases Affecting Behavior — 175
11
Sex — 193
12
Feline Parenting — 211
13
Aging — 231
14
Saying Goodbye — 247
Index
268

For my part, I am convinced that cat behavior is not only a fascinating subject to study, it is also a most promising way to probe more deeply into the evolution of mammalian behavior in general and in so doing lift the veil from our own emergent existence.

— Paul Leyhausen, *The Domestic Cat: The Biology of Its Behavior*

Understanding and Training Your Cat or Kitten

Preface

Charles F. Kettering, industrial scientist and inventor, said, "There is a great difference between knowing and understanding: You can know a lot about something and not really understand it." I hope this book will help you understand, as well as know, your feline companions. If you increase your awareness of yourself in the process, consider that a bonus.

A few years ago I decided to open a house-call veterinary service for cats, not because I had a wealth of knowledge, but because I like cats, and liking is the first step toward understanding. At the same time I wrote pet-related advice columns that appeared in several national magazines and my local newspaper. I learned about and increased my understanding of cats by observing them in their home environments, talking to pet owners who were clients, and hearing from readers of my columns.

I came to believe that cats are naturally smarter than dogs, which I treated on rare occasions. If I administered an injection to a canine patient and returned the next day for follow-up care, the dog had usually forgotten or forgiven me. There he would be, wagging his tail and seemingly glad to see me. If I saw a cat in the same situation and returned the following year, the cat was hiding under the bed or behind the couch. Cats remember!

Although cats are reported to be solitary creatures, I found that most of the pampered felines I encountered in my practice thrived on attention from their human caretakers. Many owners and cats interacted with one another through play, which included the teaching and learning of simple tricks.

One of my favorite patients was an overweight Siamese named Bubba, which lived with a littermate, Sissy, and a doting owner, Dorothy. On my first visit to Dorothy's house, she showed me a bell at the bottom of the back-door frame; Bubba swatted this bell with his paw when he wanted to go into the backyard. I marveled at both Dorothy and Bubba's ingenuity, and wondered who had trained whom.

Dr. Louis J. Camuti, who operated a house call veterinary service for cats in New York, had a theory about cat training. Dr. Camuti believed that cats trained their owners. For example, a cat may teach its owner that a twitch of the tail means "follow me" or "where's the tuna."

I think we have a lot to learn about authentic living from observing cat behavior. Cats are innately blessed with the ability to be themselves. Have you ever seen a cat that crawled into your lap and purred, just to please you? No, cats please themselves. They don't suffer from that wasted human emotion—guilt—and are uninhibited in expressing their emotions.

Although cats don't have many of the personality hang-ups we do, they are very susceptible to stress. A cat's responses to stress is directly influenced by his family history and early conditioning, plus his encounters with people, other animals, and situations in his environment.

A cat may react to stress with inappropriate or aberrant behavior, or may respond with behavior we consider weird but which is normal if you are a cat. The trick, of course, is to know the difference between normal and abnormal feline actions, and to understand the motivation behind your pet's behavior.

Motivation is the key to effective training or retraining. Castrating a tom removes his inclination to roam in search of females; changing the litter more frequently entices the fastidious cat to use his litter box, and offering a tidbit of tuna after correct performance of "fetch" answers the cat's basic question: What's in it for me?

Writing this book has increased my knowledge and understanding of patients, clients, and myself. I hope that this offering of personal experiences, my own and that of others, and research in the field of animal behavior enhances your understanding and training of the cats that share your life and home. That is what's in it for us. We're not so different from cats, after all.

1

People and Cats

The Love/Hate Relationship

Cats evoke intense emotional reactions in people, prompting wordsmiths to coin terms such as ailurophile for people who love cats and ailurophobe for those who fear them. Although I was ignorant of these words until recently, my role as a cat doctor created many opportunities to meet both cat lovers and cat haters. I use cat fearer and cat hater interchangeably here, for I believe that hate is derived from fear.

A few months after I had proudly placed the magnetic Cat Clinic signs, complete with the outline of a kitty, on the doors of the van, I found the sign on the driver's side riddled with BB-sized bullet holes; someone, a gun-toting ailurophobe I suppose, found the innocent kitty too tempting to pass up. On another occasion, I returned to the parked van to find a bearded, macho-looking individual scowling at my sign. When he realized I was the owner of the motionless kitty, he remarked, "The only good cat is a dead cat."

Was this man one of those people who made a best-seller of the cartoon book *101 Uses for a Dead Cat,* which pokes fun at cats and the people who love them? Or was he someone fed up with the ailurophiles in his life—the cat lovers who talk nauseatingly about and show photographs of their kitties just like a new grandparent or the ones who won't go to a movie after dinner because they have to get back to the cat?

I know one irate cat lover who received a copy of the dead-cat book from a friend. If the friend meant the book as a good-natured joke, it backfired. The receiver mailed a nasty letter to the book-giver and vowed never to speak to him again.

In spite of the popularity of *101 Uses for a Dead Cat,* I believe books like the best-selling *All I Ever Needed to Know I Learned from My Cat* and Lilian Jackson Braun's mystery series featuring a cat detective and Garfield paraphernalia prove that cat lovers outnumber cat haters and even those who can take or leave the so-called mysterious creatures.

Through the ages, people have had a love/hate relationship with cats. In ancient Egypt, the cat was valued for rodent control first and later envied for feminine beauty, promiscuous sexuality, mothering ability, and a mysterious, aloof nature. Egyptian ailurophiles promoted the cat as a symbol of Bastet, goddess of fertility and motherhood. As a symbol of deity, cats were protected, pampered, and mourned at death. Mummified cats, along with their bowls of milk for the afterlife, were found at archaeological excavations of Egyptian ruins.

However, it may have been the cat's association with Egyptian religion and blatant sexual nature that prompted medieval European Christians to label the creature evil and symbolic of witchcraft. As the symbol of witchcraft, the cat was banned, tortured, and killed. The persecutors perhaps brought about their own demise. When cats were killed, the rat population exploded, contributing to the Black Death epidemic.

Both in times when cats were loved and when they were hated, people had to pay a high price for going against the prevailing emotion. Humans who killed cats, even accidentally, or smuggled cats out of ancient Egypt were put to death, and those who associated with cats during the Middle Ages were deemed witches and sentenced to die.

It seems to me that the same feline characteristics that elicit love in some people produce hate in others. Strong emotions about cats reveal nothing about the nature of cats, but expose much about the person who has them.

Dr. Louis J. Camuti, in his book *All My Patients Are Under the Bed: Memoirs of a Cat Doctor,* expresses the opinion that people who hate or fear cats are those who are unsure of themselves and resent putting up with and giving in to the "significant others" in their lives. Since cats live an independent life, unresponsive for the most part to the needs of others, a defensive and/or jealous cat-hater projects his anger and fear onto the cat.

One survey investigating pet preferences revealed that 4 percent of those interviewed disliked dogs and 27 percent disliked cats. I have a theory about why some individuals who like dogs dislike cats: Cats do not engage in dominant/subordinate relationships with people the way dogs do. It is commonplace, and even desirable, for dog owners to be the top dog in the family, and the relationship is often reinforced with punishment and control. Cats, on the other hand, do not respond well to punishment and are almost impossible to control with force. An individual who likes dominance and control is going to be thwarted in his efforts with cats, for a cat will react with fear but not with obedience to these training methods.

There must be as many reasons for an emotion like love or hate as there are individuals, but it is easier for me to understand why people love cats than why others hate or fear them. Cats are adaptable and can live in a small apartment or on a large ranch. With a minimum of problems, they can live together in large cat colonies, and they can cope with, or may even prefer, life as an only-pet. Cats are clean, so you don't have to follow them around the park with a pooper-scooper or board them at the kennel while you're away for the weekend.

Individuals may be loving lap cats, more active and playful cats, or sedate and aloof cats. You can pick a cat to match your own personality.

Cats are sensitive to your moods, and give love and acceptance. They don't care if you are fat or thin, rich or poor, well or sick. But most of all, cats are soft, warm, and cuddly. It's little wonder that many people prefer to stay home with their cat rather than risk interacting with those of their own species.

Carl Jung, the famous psychoanalyst, suggested that the view one takes of animals often expresses unconscious components of the self. I suppose this means that some of us overidentify with and try to make the feline companions in our lives human, and some of us reject a fellow creature that unconsciously reminds us of ourselves. Regardless, it is hard to remain ambivalent in our attitude toward cats, and most of us divide naturally like fans of opposing football teams into those who love and those who hate cats.

Choosing Cats

Cat ownership is an investment—one requiring money, time, effort, and living space—and we should investigate the choice of cat breed, gender, size, and number, just like an investor selecting mutual funds. However, in real life, few of us—and certainly not I—are ever so organized and calculating.

I suspect that many of us don't choose a cat. The cat either chooses us or we inherit him through default. I've heard innumerable stories that go like this: I was at the flea market trying to find an antique birthing chair for Aunt Mabel's birthday, and this little tabby kitten just reached out with her paw and grasped the hem of my skirt. I heard this sound under the garbage dumpster and a calico mama with three kittens came out shivering and crying. My daughter joined the French Foreign Legion and couldn't take Fluffy, so I . . .

In these examples, we take what our heart dictates. Others choose and are also led by their heart. A scientific study showed that the appearance of the cat, primarily the color followed by the size and weight, is the most influential factor for those who pick their feline companions.

Appearance seems to be subjective with most of us. We pick the same kind of cat as our beloved cat that just died. We choose a kitty that looks just like the one we had as a kid. Or as one of my clients said, "Come to think of it, Tabetha looks just like a cat I longed for when I was twelve years old, but my mother wouldn't let me have because she cost too much."

There is something immensely satisfying about finally having what we always wanted, but the old adage about acting in haste, repenting at leisure has applicability in choosing cat companions. Physical characteristics such as gender and length of hair coat will determine, at least in part, the care needed by the cat, and factors influencing cat behavior such as personality, socialization, and activity level of the cat (to be covered more extensively in the next chapter) will have great bearing on the kind of companion he becomes.

Table 1 is a useful instrument to prompt discussion and decision among family members about expectations for cat ownership. Availability of finances, time, and facilities should also be addressed. For example, if finances are limited, selection of a "free-to-good-home" kitty, neutered and vaccinated, would be a more logical choice than an expensive, show-quality Persian.

If you own a cat-hating pit bull and the cat must share a houseboat with the ferocious canine, you might delay adopting the cat until you move to a home with a fenced backyard for the dog. If one family member is allergic to cats, another must agree to perform high-contact duties such as brushing and litter box care. If you own pets with behavioral problems, discuss the consequences of cat adoption on those problems with your pets' veterinarian or behavioral consultant.

Table 1: Considerations for Cat Selection

Use this questionnaire as a tool for discussions among family members and between family members and professionals such as veterinarians and behavioral consultants.

I. Household Members:

Name	Age	Time Spent at Home	% Responsible for Cat Care
____________	____	______	_______
____________	____	______	_______
____________	____	______	_______

How much time are household members willing to devote to the care of the cat? (estimate on weekly basis)

Time Commitment for:	Household Members			
Feeding/cleaning litter box	_____	_____	_____	_____
Grooming	_____	_____	_____	_____
Training/teaching/exercise	_____	_____	_____	_____
Arrangements for boarding, grooming, veterinary care	_____	_____	_____	_____

II. Expenses

(How much money do you estimate spending on cat care?)
Purchase price (one-time only)
Cat food (weekly)
License (yearly)
Supplies such as harness, leash, carrier, toys, bed (yearly)
Boarding/cat sitter (yearly)
Annual checkup/vaccinations (yearly, after kitten vaccinations)
Spay/neuter (one-time only)
Emergency medical treatment (yearly)

Grooming (yearly)
Pet health insurance (yearly)
Miscellaneous (yearly)

III. Living Space (Check all that apply)

Multi-family dwelling
Trailer
Farm
Quiet street
Activity/noise level high
Vacation home

Single-family house
High-rise
Houseboat
Busy street
Activity/noise level low
Nearby park/field

IV. Why Do You Want a Cat?

(Rank in order of decreasing importance)

Companionship
Rodent control
Love
Show/breed
Status

Interaction with resident pet to provide companionship/exercise
Teaching children responsibility
Tactile stimulation—petting, cuddling
Other

V. Resident Pets

List other pets already in household, specifying species, breed, age, health, activity level, and personality.

VI. Soul-searching Questions

Does anyone in the household have allergies or phobias that might worsen with the addition of a cat?

Does a resident pet have health or behavioral problems that might worsen with the addition of a cat?

What kind of cat have you always wanted and why? Does this kind of cat fit your current lifestyle, family and work situations?

Problem Relationships

Like any relationship the one we share with cats has its up and down sides. I've known those who could tolerate no down. In one case an owner turned the family cat over to the animal shelter because the cat developed a slight limp following orthopedic surgery. "I just can't bear to see him handicapped like that," said the owner. Another excuse my friends at the animal shelter hear much too often is: "The darn cat went out and got herself pregnant."

However, I've known many others who tolerate expensive, time-consuming, and heartbreaking "downs" in the relationship with their cat or cats. These owners care for aged cats, chronically ill cats, and cats with severe behavioral problems.

One Cat Clinic client quit her job so she could stay home and monitor the care of her diabetic Manx; another annually replaced the carpets in her home she shared with ten cats rather than replace the cats that urinated on the carpets. A case could be made that the love and affection shared with their pets more than compensates these cat owners for the inconvenience, expense, and discomfort the relationship brings, and therefore the relationship has more "up" to it than "down."

Investment motivates people to make the decision about what is and is not acceptable in their relationship with the pet. One survey concluded that dog owners who paid more than $100 for their pet kept the animal more than twice as long as those who received a free pet or a cheap one. It's the perceived value concept: What comes easily or cheaply is not valued as much as what is hard won.

It is probable that more cats than dogs are obtained at low cost, and those owners who fail to form a close bond with the cat are easily persuaded to let him go at the first inconvenience because their monetary and emotional investments are low.

Expectation for and control over the pet relationship also play a role in the way we perceive our pets. Drs. Eileen Karsh and Carmen Burket, who place cats with people in the Companion Cat Program at Temple University, have found three major problem areas, usually insolvable, in the development of people-cat relationships.

The first occurs when people who have not completed the grieving process replace a beloved cat that has just died with another. The new cat never measures up to the owner's expectations. The solution is to complete the grieving process and to select a new cat that is unique in its own right, not a replica of the old cat. (See Chapter 14 for a discussion about the grieving process.)

Secondly, a potential problem exists when a person who is reluctant to adopt a cat is encouraged to get one. This new cat owner may be an older, single parent who is pushed by a well-meaning but overbearing son or daughter. Since the element of choice is missing, the person/cat relationship is never cemented. Another example is that of the overworked mom who is harnessed into getting a cat by her young, vocal offspring. In some cases, Mom would be better off with a hardy ivy plant than a long-haired Persian requiring daily combing and brushing.

Problems also arise when the personality of the owner is mismatched with that of the pet. An example would be placing a highly active and aggressive cat with a frail, elderly owner.

Another area of difficulty between people and cats occurs when people redirect their feelings for other humans onto the cat. A man or woman who is angry or upset with his or her spouse may direct that emotion toward the cat rather than the spouse. It's the old "kick the cat" syndrome.

Family members' personalities influence the selection of a cat that will fit into the family in a complementary rather than an antagonistic manner. The emotional traits of felines must be known or investigated before matching cats with people.

Table 2: Matching Feline and Human Personality Types

(A) Sociable, confident, easygoing cat: This type of cat personality blends well with most owners and families. A relationship with a friendly cat may even be therapeutic for the mildly depressed owner. This cat can hold his own in a household of noisy children.
(B) Timid, nervous, shy, unfriendly cat: This cat personality is a poor match for an owner who is underconfident and suffers from poor self-esteem. The owner often perceives the cat's actions, such as hiding, as a personal rejection. A nurturing, patient owner can gain this cat's loyalty. This cat is intimidated by active children.
(C) Active, aggressive cat: Does not form close attachments to people. Poor choice for elderly owners or families with small children.

The Bond

Feelings of love and concern for a person can be redirected toward the pet. One story that always brings tears to my eyes is that of the man who cared for a Siamese belonging to his son, who was killed in Vietnam. All of the love and devotion that could not be given the son were lavished on the son's cat. When the cat expired, the grieving father was devastated, more so than when the son died, some people said.

Redirected love is just one of the ways a bond is developed between people and pets. People develop a strong attachment to a cat by caring for it. Just like a child, the cat becomes dependent upon his owner. I found that I often thought of my cat C.K. as one of the kids and responded to him in the same tone of voice I did for Kimberly, my youngest daughter. "Get off the table, C.K.!" was the same as "Kimberly, get in there and clean your room!" I even went so far as to occasionally call one of them by the other's name.

Dr. Victoria Voith, a behavior specialist formerly at the Veterinary Hospital of the University of Pennsylvania, compiled a questionnaire about cat bonding to which 872 owners responded. Ninety-nine percent considered their cats family members, and the majority used the word "love" to describe how they felt about their cats. Over 90 percent of owners talked to their cats daily and 89 percent reported that the cats slept on their bed.

The benefits of a strong bond between people and cats have been demonstrated by many studies. In one investigation conducted in 1983, elderly cat owners who developed a long-term relationship with their cat were compared with owners who gave up their cat within a year and those who did not adopt a cat. There were striking differences in feelings of well-being. Long-term owners felt better physically and were less depressed mentally and less lonely.

The benefits are not restricted to elderly owners. Young, single adults who are cat owners are not only less lonely, they also often talk about their cat being the catalyst for interactions with other people.

Two of my clients, each owning five cats, met during my tutorage as their cat's doctor. The two cat lovers found they had much in common, and I delighted in monitoring the progress of their romance as I monitored the health of their cats. The lovers married, blending two cat households, and I gave them a pair of cat-shaped bookends for a wedding present.

The feelings people have for their pets are demonstrated almost daily by articles in my local newspaper. In today's Ann Landers column, the following letter expresses the sentiment of people writing about their pets:

> I cannot imagine my life without a dog or cat. My growing-up years were enriched a thousandfold by Muffie and Tiger, the two cats that slept on my pillow at night. . . . Every year when I write checks for my favorite charities, our local animal shelter is at the top of the list. My donations are always "In loving memory of Muffie, Tiger and Buster."

This letter writer is not alone in intoning the blessing of cat ownership. Testimonials abound of cats that spark communication with depressed or autistic children, compassion and rehabilitation in prisoners, stimulation for lackadaisical nursing home residents, and healing for physically or mentally ill patients, and cats that serve as hearing/ear cats or save the family from a burning house.

For most of us, cats add excitement and quality to life, but for some, cats give meaning to life itself.

Questions

Dear Dr. Whiteley,

My six-year-old son has allergies. His physician thinks that the two family cats are the source of some of the allergic symptoms and recommends that we get rid of them. This is causing great dissension among family members. My husband says the cats have to go, while my daughter says she would rather get rid of her brother than part with the cats. I'm so tired of all the bickering that I'm ready to keep the cats and get rid of my husband and both kids. Do you have any suggestions?

Fed-up in Dayton

Dear Fed-up,

I know how you feel. I think about sending my husband and kids to the pound at least monthly.

Most allergic people are able to find alternatives to giving up their pets. I'd like to make a few suggestions that include allergy-proofing cats and house.

Brush the cats daily with a soft brush to remove allergy-causing dander (hair and dandruff) and saliva. Wipe the cats daily with a towel or sponge moistened with distilled water.

A relatively new product called Allerpet (marketed by Veterinary Products Laboratories through veterinarians and pet stores) is designed to reduce the allergens in dander and cat saliva. You wipe the cats' fur with a damp sponge saturated with Allerpet once weekly; when using the product, do not apply the daily wet towel treatment.

Shampoo cats every two weeks with a mild pet shampoo and creme rinse. Feed the cats a balanced diet containing natural fat. If the cats suffer from dryness of hair, ask your veterinarian about adding fatty acid supplements to their diet.

Have a nonallergic family member (I vote for your daughter) take care of the catboxes at a time when your son is elsewhere. Use a dust- and perfume-free litter product, and take care of the boxes daily.

Allergic persons are especially sensitive to odors; therefore, neuter males to eliminate tomcat odor. Remove pet food dishes and wash them after feeding time. Clean cat beds often, and avoid the use of insecticides and perfumed products on the cats. Your son may be sensitive to cat scratches; if this is a problem, have the cats declawed.

Most people who are allergic to cats are also allergic to dust, pollen, and mold. Allergy-proof your home by installing an industrial-sized air purifier, and use air-conditioning and humidifiers in season. Forbid smoking in your house, and vacuum often when the allergic person is away from home. Eliminate dust-catchers such as dried-flower arrangements and blinds, and sources of mold in damp areas such as bathrooms and basements. Use nonallergic, polyester-filled pillows and comforters.

On a positive note, with time, some people seem to develop a tolerance to their own cats. Allergic children often discover that allergy symptoms decrease as they grow older. Desensitization (allergy shots) may benefit certain allergy sufferers.

Best wishes for finding solutions that will enable you to keep both family members and cats happy and healthy.

H.E.W.

Dear Dr. Whiteley,

My wife, Doris, and I have been married two years, and I doubt we'll make it together another two. The problem is Doris's seven cats. The cats consume all of my wife's time and interest. She recently quit her part-time job to stay at home with the two older cats, both of which have chronic medical problems.

Doris refuses to go out of town for a vacation because she doesn't want to leave the cats. We no longer have a social life; the cats get nervous if people come over, and my wife gets nervous if she leaves the cats alone in the house. There is no hope of the cats dying and leaving us in peace, for Doris would just adopt others. Frankly, I'm ready to throw in the pooper-scooper and call the marriage quits.

Cat-pecked in Sacramento

Dear Cat-pecked,

My heart goes out to you. It is hard to compete with seven furry balls with big eyes that evoke neediness. Playing caretaker to seven needy felines is a role suited to a codependent cat lover such as your wife.

"Codependent," in pop psychology circles, refers to a person with dependency issues. The cats are dependent upon your wife, but your wife, in this case, is also dependent upon the cats. She probably feels responsible for their happiness and well-being. She may have rescued some of them from terrible fates and feels she must continue to rescue them from the stresses of old age, chronic illness, and strangers.

When being the care-giver to the cats takes priority over your wife's needs, your needs, and marital needs, it becomes a sickness. Like any other illness, it must be recognized to be treated. It won't be until your wife lets go of her intense feelings of responsibility for the cats that such suggestions as hiring capable cat sitters, counter-conditioning the cats to accept strangers, and discussing the older cats' health care with your veterinarian will be heeded.

Ann Landers would, I'm sure, suggest marital counseling. Along with counseling, a codepedency support group for you and Doris would be helpful. If you cannot find such a group in your telephone yellow pages, consult your local Al-Anon chapter for a referral. Read Melody Beattie's book, *Codependent No More.* I discovered that most of the characteristics of codependency suggested by Beattie applied to me. Caring veterinarians and cat lovers are susceptible to letting our feelings for our feline charges go too far until we slip over the edge into codependency.

Good luck!

H.E.W.

2

Starting Right – Kitten or New Cat

Family History

Are we what we are because of family history or in spite of it? My opinion is that we, human and feline, are a combination of everything present and past, including Adam and Eve and the first cat that pounced on the serpent, while remaining individuals who are different and unique from all others, regardless of family history.

Cats become what they are genetically and physically predestined to be, yet some natural propensities can be enhanced or overridden by correct training and socialization.

Selecting a Kitten or Cat

Genetics and Breeding

Kittens show the personality traits of their father, even when the offspring never come into contact with him. Kittens also inherit the mother's traits, but it is difficult to determine if her influence is genetic or the result of her association with and training of the kittens.

Those who are in the position of selecting their new kitten before birth would be wise to investigate the health and behavioral traits of prospective feline parents. One cat breeder might be especially proud of the amiable personality bent of his felines' family tree and readily share that information with you, while another may try to overlook and hide the fact that your kitten-to-be's mother has bitten and scratched four judges and two bystanders unfortunate enough to stand too near. It is up to you to search out the truth.

Certain feline breeds are known for congeniality more than others. For example, the Siamese is considered gregarious and friendly, while the Rex is more nervous and withdrawn. Of course, many of the cats people adopt, and all the ones I have owned, are more generic than blue-blood as far as ancestry goes. In that case, you have to guess at personality tendencies by observing the kitten's parents.

Table 3: Personality Traits of Common Cat Breeds

Abyssinian
shy
tends to be fearful of strangers
nervous
intelligent
loyal to owner

Burmese
active
wants attention
playful
easygoing

Himalayan
outgoing
friendly
moderately affectionate

Persian
reserved
inactive
shy
quiet

Rex
nervous
withdrawn
fearful

Russian Blue
quiet
shy
withdrawn

Siamese
outgoing
intelligent
demands attention
loud

Physical Characteristics

Physical characteristics also affect behavior. For instance, cats with heavy, thick coats tend to shy away from excessive cuddling because they become too warm.

Cats with handicaps react differently from those without them. Blue-eyed white cats and, to a lesser extent, white cats with orange eyes are born deaf. The deafness is linked to the gene producing white hair color. People often perceive these cats as mentally slow when in reality the slowness is due to an inability to hear.

Gender

Another consideration during the selection process is gender, and both males and females have their advantages and disadvantages. Males are generally larger than females and in the wild enjoy a larger home range, factors that might influence an owner living in a tiny efficiency apartment. Male cats are more likely than females to spray urine, a decidedly unpleasant behavior, and to fight with other males. On the other hand, females exhibit loud and occasionally unruly behavior associated with heat unless neutered.

I was guilty of recommending females over males until one day I remembered that my own favorite cats have all been males. Therefore, I now refrain from offering an opinion about gender selection except in cases where a client is selecting a new cat to join a resident cat that already has behavioral problems which might worsen by the addition of one sex or the other.

Age

What are the differences between adopting an adult cat and a kitten? It is easier to know what you are getting with an adult cat; his personality, habits, and physical characteristics are already formed. Kittens might be considered pure potential. A kitten's behavior is more easily molded by training than a cat's. Kittens are usually active, curious, precocious, and fun or a "pain-in-the-tush," depending on your point of view.

My grandcat, Prissy, age four months, spent the Thanksgiving holidays with my husband and me while her owner, our daughter, was away. I had forgotten how busy a kitten can be. Prissy sprang out from behind the end tables to attack our shoelaces; scattered the stacked bath towels, still warm and fluffy from the dryer, into rumpled disarray; and slept on top of our heads the first night. Finally, we locked her in the utility room with her own bed. Some of us love these stimulating interminglings, and others of us worry that we'll fail grandparenthood when we try it again with a two-year-old toddler.

When Two Is Better Than One

Cats, as well as people, vary in their desire for interaction with their own kind, and a need for social intercourse with other cats is influenced by the cat's upbringing, which will be discussed more fully later in this chapter. If you are an owner who will be away from home much of the time and want to adopt two cats that can socialize in your absence, I recommend that you adopt two young kitties, littermates preferably, that can form social bonds with each other as they grow up. The desirability of adopting another cat or kitten as playmate for your resident cat will depend upon the social nature of your cat.

Deprived Kittens

Prospective cat owners should be aware that the physical health of the mother cat affects the mental health and personality traits, as well as physical health, of her litter. A cat mother unable to provide milk for all the offspring of a large litter may contribute to the mental and physical retardation of a weak kitten unable to compete for his nutrients. Kittens orphaned soon after birth may experience a similar lack of nourishment as well as nurturance.

Deprived kittens often experience delays in behavioral development, including crawling, suckling, eye opening, grooming, walking, running, playing, and climbing. Kittens of undernourished mothers are slow learners. In spite of adequate nourishment provided after birth, kittens deprived of nutrients during critical development in the mother's uterus do not catch up with kittens that receive adequate prenatal care.

This does not necessarily mean that a retarded cat will be unable to function as a pet. I am reminded of Blackie, a kitten two sobbing twelve-year-old girls found in the street in front of the veterinary hospital where I worked. Blackie's tail had been amputated by a car.

I consoled the young Samaritans, surgically smoothed over Blackie's remaining tail nub, and took custody of the cat. My daughters were young and playful then, and poor Blackie was immediately pressed into service as a compliant playmate. I'm sure that a more intelligent animal would have made himself scarce after the first encounter with the girls.

Blackie, however, never learned to hide or run when his human tormentors were around and would sit for hours in a doll carriage, dressed in uncomfortable baby clothes, with a dumb look on his face. I suppose I should be ashamed of having allowed what some would call cat abuse, but at the time I was grateful for any diversion that kept the girls occupied. I was also appreciative of Blackie's tolerant, if not so bright nature and considered him a highly desirable pet.

The First Weeks

Handling and Other Stimulation

Providing mental and physical stimulation to a newborn kitten gives it the start it needs to develop into a supercat. Young children, with proper adult supervision, should be allowed to play for limited periods with feline babies if the feline mother permits it. An added benefit is that the kitties learn to tolerate and enjoy all kinds of interactions with kids. In one study, kittens handled for ten minutes a day starting just after birth opened their eyes a day sooner than those not handled. The handled kittens emerged from the nest three days earlier and were more active than those not touched.

Kittens in cluttered rooms receive more stimulation than those raised in neat, tidy rooms. All those paper bags and boxes on the floor provide objects for play and interaction between siblings. What an excuse for teenagers who keep a messy room: It's okay, Mom, I'm providing learning opportunities for Miss Kitty.

Weaning

Ideally, kittens should stay with the litter, and weaning should commence no earlier than four weeks or later than six weeks; eight weeks is the preferable age for most kitten adoptions.

Of course, not all adoptions occur ideally, and I'm not recommending that you give back to nature the orphan kitty that claimed you for his parent one rainy night when he was two weeks old or the one that came to you belatedly as an older cat. However, it helps to know what to expect, I think, so you can understand and work with your cat to overcome negative conditioning.

Isolated Kittens

Kittens isolated from their primary associates—mother, siblings, and owner—at two weeks of age often become withdrawn and fearful, reacting aggressively toward other cats and people. Novelty or unusual circumstances put an enormous amount of stress on an individual weaned early.

Male kittens weaned early and raised in isolation develop poor grooming habits and may have a difficult time, when grown, with sexual performance. These are the macho males seen dragging the female around by the scruff but not actually impregnating her.

Kittens raised as an only-child often manage to delay weaning. They continue to nurse, if the poor mother is willing, until they are bigger than mom. Many of these kittens grow up to exhibit separation anxiety, which may escalate into destructive behavior, when their owner, the new mother figure, leaves them alone.

Kitten's Work—Play and Hunting School

Kittens learn to be cats by exposure to and interaction with mother and littermates. Kittens raised without mother and littermates do not learn appropriate social communication skills. As cats mature, they continue to learn social rules from their owners and other animal playmates.

Play

It is said that play is the work of children. The same is true for kittens. Through play with siblings, kittens learn to control how hard they bite and to retract their claws to prevent injury to their playmates.

Broad categories of play include play-fighting such as wrestling and chasing; play with objects, real or imaginary; and locomotion play, such as leaps and acrobatics. Play also resembles the stalking, chasing, and pouncing behavior necessary for catching prey.

Hunting

Although the physical movements necessary for rodent hunting are inborn in kittens, evidence shows that hunting proficiency is learned. How do kittens learn to be mousers? They observe the mother: Mother carries home prey that she has killed and eats it in the presence of her kittens, mother next leaves dead prey for her kittens to eat, and finally mother brings home live prey and allows the kittens to catch and eat it.

In one study, eighteen kittens raised alone with rats or mice for littermates refrained from killing their rodent cagemates or even other rats or mice when they had the chance. Hunger and observing other cats killing rodents had no effect on their behavior. These kittens played with the rodents just as they would with other kittens; they protected their cagemates from intruders, and were observed meowing, searching, and moving restlessly about the cage when the rodents were removed. Kittens raised with another kitten did not form attachments to rodents.

This leads to the question of whether or not you want your kitten to be a mouser. If you want a cat for rodent control, select one whose mother is a good mouser; mother will do the teaching for you. If the sight of a cat with a mouse or small bird in his mouth horrifies you, you would do well to adopt a kitten from a mother that has never been exposed to rodent killing.

Socialization

Socialization is the process whereby cats learn to relate to others, human and animal, in their group. Cats raised in the wild tend to be asocial, and will leave people, other cats, and dogs alone. Most of us adopt a kitten or cat to provide companionship, and prefer a pet that at least acknowledges us and preferably comes to regard us as a friend. Socialization is how you teach your kitten or cat to be congenial, and the process can be approached in a pleasurable and fun way.

Play and Toys

Playing with your new kitten or cat not only stimulates learning for the feline but also helps socialize your pet to the human race in general and to you in particular. This attachment process is especially important during the weeks before and after weaning.

Choose games and toys that are fun, safe, and stimulating to you and your kitten or cat. Dr. Gary Guyot, a psychology professor at Regis College in Denver who has studied play behavior in cats, suggests taking your cue from the cat: "Observe what the cat likes to do and how much he wants to play. Cats tend to like things that are smaller than they are; a small rubber ball provides more enjoyment than a large beach ball."

Cats are adapted to see movement. The horizontal movement of a piece of string dragged in front of the cat or a rolling ball proves enticing to most cats.

Get down on your kitten or cat's level when introducing a new game; it's hard for a half-pound kitten, or even a fifteen-pound fat cat, to relate to a towering giant. Because kittens become excited and use their teeth and claws in play, it is rarely a good idea to interact with them using your bare hands. Use the pipe-cleaner toy or crumpled paper ball to make contact and don't instigate tug-of-war or other aggressive games.

Provide a variety of toys and games, and rotate your pet's toys once or twice a week. If your cat enjoys catnip, rub it on old toys to give them fresh appeal. Offer string or yarn as playthings only when you are there to supervise. If ingested, these items can cause intestinal problems. Beware of toys that contain squeakers, buttons, or other small parts that can be chewed off and swallowed.

Teaching your kitten simple tricks is also mentally stimulating and helps your feline bond with you. Tricks will be covered more extensively in Chapter 7.

Attachment Period

Let's go back a moment to that sensitive attachment or socialization period, occurring for the most part before and during weaning. Research suggests that what the kitten is exposed to during this time imprints in such a way that the cat carries its impressions throughout life. Attachments and impressions are formed easily and fairly rapidly. In the same way that a kitten exposed to lots of noise and children will grow up to tolerate, and probably enjoy, children, a kitten experiencing frightful situations, a thunderstorm, for example, may develop phobias such as fear of storms.

Since this is the time that most kittens meet the veterinarian and travel in the car for the first time, you should be careful that these prove to be pleasurable experiences. If you, the kitten's new mother, act fearful and say in a loud voice, "Oh no, that mean old doctor is going to give my poor baby a shot!" the cat may not understand your words, but he will get the message and act accordingly. I have scars from encounters with cats trained, albeit unintentionally, by owners to fear veterinarians.

Exposure to Humans

Recent research suggests that the prime time for socializing kittens to people is between ages two and seven weeks, earlier than for puppies. This means that most people who adopt kittens must depend upon the owners of the mother cat to socialize the kittens to people. Attachments and preferences can be changed later, but it is a more tedious process.

Both feline and human mothers facilitate the attachment of kittens to people. If the cat mother is distrustful of people and hides the kittens after two weeks of age, the process of socializing kitties to people is difficult. If the cat mother brings her kittens into the human social circle, interacting with both people and kitties, her offspring seem to attach naturally to people.

It appears that when it comes to the human handling of kittens, more is better. In the laboratory, anyway, kittens that were handled forty minutes a day became more people-oriented than kittens handled only fifteen minutes. Although the act of feeding a kitten enhances his relationship with the person giving nourishment, feeding alone is not sufficient to maintain the relationship. It takes other interactions such as petting, playing, and talking with new kittens to really solidify the affiliation. Of course, who can resist petting and talking "kitty talk" to a new cat baby.

INTRODUCING STRANGERS—While operating my cat clinic I met many cats that had bonded well with their owners but feared strangers, and some had specific aversions to a race or sex different from that of the owner. This has the potential to be traumatic when a solitary owner suddenly changes lifestyles or adds family members. One example that comes to mind is Misty, a three-year-old calico cat raised by a young, single woman named Roberta.

Everything was fine between Misty and Roberta until Roberta started a serious courtship with William. Whenever William visited Roberta's apartment, Misty spent at least an hour hissing and howling at him. Then the cat ran under the bed and refused Roberta's coaxing to come out of hiding. If William phoned, Misty would try to knock the telephone receiver out of Roberta's hand. It soon became an either/or situation: Either Misty changed her attitude and behavior or Roberta had to choose between her boyfriend and her cat.

If Misty had been introduced to men before or soon after adoption, she would have reacted more favorably toward William. However, when Roberta adopted Misty, the young woman had no men in her life, and that was one of the reasons she chose a feline companion. If Roberta had known about the socialization period, she could have taken steps to familiarize her cat with the human male during that sensitive period in Misty's life. Those steps might have included playing recordings of men's voices, opening a bottle of men's aftershave, and inviting the mailman in for coffee. Who knows what else she might have accomplished. I don't want to leave you worrying about Roberta and William, if you are a romantic, or Misty, if you are an animal lover; suggestions for correcting Misty's misbehavior will be given in Chapter 8.

INTRODUCING BABIES–Feline jealousy also can become a problem when an only-cat is suddenly forced to share love, attention, and space with a newborn baby. If you know in advance that you will be starting a family, introduce your kitten to babies—and make sure it is a pleasurable experience—soon after adopting the pet. If, however, the baby-to-be is a surprise, you can make your feline's adjustment easier by taking certain steps:

Expose the cat to the sounds and smells of a baby: baby powder, baby oil, diapers, recordings of a baby crying. Borrow a baby from a friend; most new parents are more than willing to give up Junior for a couple of hours while they take a nap. Let your cat see and hear you cuddling and cooing with the infant. At the same time, have your spouse hold and cuddle the cat. Change places with your spouse—you hold and cuddle the cat while your spouse interacts with the baby. The goal is to have the cat associate the baby with more, not less, attention from family members.

After your own baby arrives, try to spend quality time, even if for a short period, with your cat, and don't leave baby and cat together unsupervised.

Exposure to Animals

For those of you who have the opportunity to train your kitten during the socialization period, this is the time to introduce dogs and birds or other animals, if your pet is to share quarters with and bond with these species. This bonding between animals is demonstrated by the story of KoKo, a captive gorilla accomplished in sign language that had adopted a pet kitten. National magazines printed touching photos of the large and bulky KoKo gently cuddling her tiny feline friend. Then the kitten died, and KoKo grieved. The lonely gorilla cried and signed that she missed her kitten. I'm sure the kitten would have grieved, also, if KoKo had died first.

INTRODUCING RESIDENT CAT TO NEW CAT–If your cat is older and you decide to add another cat, your resident cat is likely to perceive the new cat as an intruder in his territory and react with hissing and other aggressive behavior; you will have to proceed slowly and cautiously when making these feline introductions.

If possible, allow the resident cat to become familiar with the new cat's scent before bringing the new kitty home. This can be accomplished by placing the new kitty's blanket or bed or, even better, a cat carrier containing his bed in an area where the resident cat can sniff and become acquainted with it for several days.

When you arrive with the new cat, leave him in the cat carrier while you make the introductions. Allow the two to get used to each other for a short period—twenty or thirty minutes—and then place the new cat in a separate closed room. Repeat the visits between the two cats several times a day, including mealtime, with the new cat remaining in the carrier. When feeding the cats, place the respective bowls together, with the carrier door as separation, so that the two cats can face each other without competing for food.

After the cats have become familiarized, allow them to interact without the carrier when you are present for supervision. Again, it might be wise to limit this time until you know the two can be trusted together for longer periods, which will vary, depending upon the reaction of one or both cats. Provide each animal with separate feeding and water bowls, beds, and litter boxes.

INTRODUCING RESIDENT CAT TO NEW DOG–Many of the same principles apply when introducing a new dog to a resident cat: Proceed slowly, do not initially allow the two animals contact without supervision, and provide separate feeding, sleeping, and eliminating areas. Plus, make sure your new dog is obedience trained or enroll him immediately so that he can be brought under voice commands when interacting with kitty.

Exposure to Life Situations

Socialization of your kitten includes exposing the little fellow to as many things and situations as possible after he has adjusted to family members. This includes trips in the car, water for baths, and toilette training, which will be addressed specifically in Chapter 4.

Everything you do is sending a message to your kitten or cat. Make sure it is the message you intend, be consistent, and have fun.

Question

Dear Dr. Whiteley,

I want to know how to determine the sex of newborn kittens. The last time my mother cat had a litter, I had no trouble giving the three babies away. However, two of the new owners called back to complain. It seems that George should have been named Georgette and Princess should have been Prince.

Confused in Detroit

Dear Confused,

If any of the newborns are tricolored (calico or tortoiseshell), give them feminine names; 98 percent of cats with this coloration are female. You can determine the sex of the remaining kittens by comparing the distance between the anus and genital opening. The ano-genital length will be greater in males to accommodate for the scrotum and testicles that will become more visible as the little fellows mature. If the length is the same for all kittens, gently expose the genital opening of one of the kittens, examining for a penis if the kitten is male or for lips of the vulva if female. You might also consider unisex names like Blackie, Kitty, or Fluffy.

Good luck!

H.E.W.

3

Sensing, Learning, and Communicating

Cat Sense

Hearing

Cats are creatures of their senses, especially those of smell and hearing. Feline hearing is more sensitive than human hearing, with an upper hearing range of 50,000 to 60,000 waves (Hz) per second (ours are at 18,000 to 20,000 Hz).

In general, smaller animals hear higher frequencies than larger animals. The higher the pitch of a sound, the easier it is to locate its origin. The mouse has an extremely high-pitched voice, reaching ultrasonic frequencies at its peak. A vocal mouse is inviting attention from an alert cat.

The cat's cup-shaped outer ear rotates 180 degrees and serves as a funnel, also enabling the cat to pinpoint a sound's source and to amplify the sound's intensity.

Certain high-frequency sounds inaudible to us, such as electronic pest-control collars or remote-control devices used to change television channels, may be disturbing to a cat. If your pet expresses anxiety by moving away or vocally protesting when you activate one of these modern electronic gadgets, be aware that you may be sacrificing your pet's comfort for yours.

Even when cat napping, which is how most cats spend the majority of their time, the feline's radarlike senses remain on duty. The sleeping cat is able to sort through tremendous amounts of irrelevant noise and pick out significant sounds. If he hears a strange noise, the cat briefly assesses it, and if the sound presents no danger or intrigue, he will ignore it. If, however, the new sound is that of you operating an electric can opener, the little fellow that was comotose seconds before will be at your feet to see if you are opening a can of tuna or sardines.

Vision

It is estimated that a kitten's ocular development at birth is similar to that of a five-month-old unborn human fetus; the kitten's eyes are sealed shut at birth and do not open for several days.

Visual acuity develops independently of eye opening. Even after opening the eyes, kittens must use them for vision in order for critical neural pathways between the eyes and brain to develop. Kittens raised entirely in the dark will act as if they are blind, and kittens raised in a room painted solely with black-and-white vertical stripes will have difficulty perceiving horizontal shapes.

Vision is important in developing certain aspects of locomotion and depth perception. Blind kittens are going to walk, run, and climb later than their sighted siblings, and experience difficulty with foot placement and normal interactions with companions and environment.

By two months of age, a kitten raised with normal visual stimulation has attained the sight capabilities of an adult cat.

Cats have the largest eyes in relation to their head size of domestic animals, an attribute contributing to their beauty, charm, and visual powers. Because of eye position (in front and facing forward) and head shape, the cat has an overlapping field of vision (binocular vision) just as we do. The overlapping fields of vision help us judge distance, pick out details, and see when light is poor.

In contrast, prey animals such as rabbits have an eye on either side of the head. This enables the hunted to see as much of their surroundings as possible, so they will not be caught unaware by predators.

THIRD EYELID–The cat has a third eyelid, missing in humans, which functions to move tears across the eye and to protect the eye from injury. Sometimes, this white membrane will almost completely cover one or both eyes, causing concern in cat owners.

Cats that receive much tactile stimulation, such as fondling, may show persistent exposure of third eyelids. It is believed this is due to stimulation of the nervous system. Temporary separation of owner and cat is usually curative.

Protrusion of the third eyelid is seen occasionally in healthy cats but may also be a warning sign of eye irritation or disease. Consult your cat's veterinarian if the cat's third eyelids persistently shield his eyes.

NIGHT VISION–A cat's eyes appear to glow in the dark because of the tapetum lucidum, a layer in the back of the eye that reflects light and allows cats to use more available light for activities like night hunting.

Cats see well at light levels ten times lower than humans can see. The feline pupil opens wider, and the retina has more light-gathering rods than ours does. Since cats function well at low light levels, they often appreciate a quiet, darkened room, especially when stressed or recuperating from illness, or as a preferred location for their private toilette, the litter box.

Because of the vertical arrangement of the cat's pupil and the horizontal arrangement of his eyelids, this anmal can deliberately close off the light entering his eyes; rarely is a cat blinded by approaching car headlights. This may be one reason that fewer cats than dogs are injured by automobiles.

DETAIL AND COLOR–Cats are nearsighted and see in muted shades, much the way objects look to us at dusk, when most cats like to go out hunting or carousing. As the cat moves closer to an object, his ability to see sharper colors is enhanced.

Cats don't see detail well, but they are great at observing horizontal movement, which is how most of their prey move. A string pulled in a horizontal direction or a small ball rolled along the floor offers much more inducement for the cat to go for it than if the same objects were moved up and down.

Touch

The cat's sense of touch is enhanced by his whiskers, located on his upper lips, above his eyes, below his ears, on his chin, and behind his front paws. The small tuft of foreleg whiskers increases his sense of feel for catching mice or other prey. His facial whiskers face forward and to the side, and serve to scan a wide area when the cat walks; the cat's whiskers fold against the side of his face at nap time.

A cat without whiskers is unsure in the dark and has difficulty negotiating narrow spaces. The information from the whiskers seems to act in concert with visual information, allowing the cat to form a three-dimensional view of his surroundings.

HEAT–Cats do not seem as sensitive to hot objects as we do, which may be one reason a cat may sit nonchalantly on a hot radiator until his hair begins to singe.

Cats are also attracted to warm objects when it is cold. Winter brings an outbreak of cats injured because they sleep undetected under car hoods in order to be closer to warm auto engines. An unsuspecting cat owner starts the car, and the car's fan belt makes mincemeat of the cat.

Infant kitties are unable to maintain their own body heat, so they use their sense of touch to push against warm objects, especially mother and littermates. Physical contact with mother stimulates the young to bury into the older cat's hair, and a good mother accommodates her babies by curling her body around them as they nurse and sleep. When raising orphans, it is a good idea to provide a warm object such as a hot water bottle covered with soft material for the babies to snuggle against. Providing something soft, warm, and snuggly to rest upon is also good nursing for a sick or injured individual.

Sometimes, an apprehensive cat can be calmed by having a beloved owner place his or her warm hands over the cat's face; this action reminds the feline patient of that early nurturing, I suppose.

Smell

Cats have a very good sense of smell. The cat's nose and nasal passages are larger than the corresponding area in humans, and that part of the feline brain responding to scent contains about 67 million cells, 1.5 million more than ours.

AGREEABLE/DISAGREEABLE AROMAS—Fragrances that are exotic and stimulating to a cat may smell terrible to you and me. Cats like to eat what smells good to them, and the pet food industry has capitalized on this propensity. A liquid made from the fermentation of ground poultry organs, fish, liver, and beef lungs is often sprayed on dry cat food during processing to increase its appeal to cats. Cats love the smell of fermented liver and lungs.

Cats react to odor. Occasionally, a feline behavioral problem, such as eliminating outside the litter box, occurs when the cat detects or is suspicious of a new litter containing a deodorant that appeals to the owner but not to the cat. Sometimes, a cat will refuse to come to the outstretched hand of his favorite person if that hand has recently dug into the moth crystals during spring cleaning. On a more positive note, sick cats can often be enticed to eat when their food is heated to volatilize its odors.

SCENT MARKING–Cats are very conscious of their environment and take great pains to leave their scent signature to prove their right to it. It's a way of saying to other cats: Hey, take notice. Fluffy has been here.

One way cats mark is with urine. A female in heat signals the males in her locale that she is ready and waiting. A tom spraying urine may be warning away contenders, and a neutered male may be claiming territorial rights.

The sweat glands in the footpads also leave the cat's individual scent on a favorite tree, a scratching post, or a velvet couch, depending upon his training. Damp paw prints left on a veterinary exam table usually mean the cat is nervous, however, rather than staking a claim to the hated table.

SCENT GREETING–Cats use their scent to assert ownership. When your cat is vigorously rubbing his head and body against your legs, he is leaving his scent from sebaceous glands located primarily in his lips and chin so other cats will know that you are his human.

In a similar way, cats greet other felines they know. First they greet each other face-to-face and then present their anus or rear for the other cat to smell. You may not appreciate it, but your cat often presents his rear for you to smell as a gesture of his esteem for you.

If you are introducing yourelf to a new cat, it is a good idea to proceed slowly just as another cat would do. Talk calmly and softly, avoid staring directly at the cat without blinking, and get down on the cat's level. Cautiously, offer a palm turned upward for the cat to smell. The cat's body language, to be covered later in this chapter, will be the clue to whether it is safe to pet or pick up the animal or it would be prudent to make a hasty exit. Regardless, you will have used proper cat protocol in making your introduction.

VOMERONASAL ORGAN–Cats have a secondary scent retrieval system in the roof of their mouth called the vomeronasal organ. When the cat approaches something with an odor he wants to experience more fully, he activates this sensitive tissue by inhaling through a partially open mouth. The cat often appears to be grinning, but he is actually appreciating what to him must be a fine perfume. This gesture, called the Flehmen, appears most often on the face of a tomcat smelling the genitals of a female in heat.

It is through scent, as well as touch, that a newborn is attracted to the nipple, a preferred nipple he tries to retain over the duration of nursing. As stated, a cat uses his sense of smell to establish territory, attract a mate, and select dinner. I've known blind and deaf cats that seem to live a rich life, but a cat deprived of his sense of smell would be handicapped indeed.

Taste

The domestic cat has fifteen hundred taste buds occurring in groups within tiny fingerlike projections that can be seen by looking closely at the cat's tongue.

Taste, especially in the cat, is intimately associated with sense of smell. What smells good usually tastes good, as discussed in the example of the fermented liver and lungs (page 37).

Although cats can develop a strong liking for sweet foods, leading to all the problems we know so well, such as tooth and gum disease and obesity, it is usually not a feline taste preference. When selecting food treats for your cat, avoid sweets. It's unkind to the cat to encourage him to develop a sweet tooth.

Psychic Sense

Are the physical senses of hearing, sight, touch, smell, and taste all? What about cats that predict earthquakes or erupting volcanoes by packing up their kittens and getting out of town before the disaster occurs?

How do these cats perceive an earthquake or erupting volcano? I don't think anyone knows for sure, or we'd use that knowledge to improve our own predictions of natural disasters. It is surmised, however, that certain animals are exquisitely sensitive to minute changes in electrical charges in the air and hence know danger is near.

I have always been fascinated by accounts of animals traveling hundreds of miles across mountains and interstates to come home.

Vincent and Margaret Gaddis recount an interesting story in their book *The Strange World of Animals and Pets* about a couple who had left Oklahoma with their cat Sugar for the bright lights of California. Later, the displaced Okies decided to return home to the farm in Oklahoma, and left Sugar with a friend in California. Fourteen months later, bedraggled Sugar showed up at the Oklahoma farm just in time for milking.

How did Sugar know that her human companions had moved back to the farm in Oklahoma? Did she go back because she was looking for her people or for a remembered place? How did she travel over a thousand miles to the correct destination? Other cats become disoriented and lost when they find themselves in a strange place. Did she hitchhike with a sympathetic trucker driving an eighteen-wheeler with an Oklahoma license plate? I'd really like to know the answers to these questions.

I've found some explanations by reading the scientific literature, but not the answers to these specific questions. For one, the homing instinct is not learned. Young deer mice with no experience will return to the home nest. Other animals that home take the shortest route, not one they've had experience with. And, as previously mentioned, some individuals are better at it than others.

Surely, the unique and sensitive felines traveling hundreds of miles home without benefit of road map, those predicting natural disasters without benefit of sophisticated equipment, and those expressing a knowingness or sadness when a beloved master dies many miles away in a hospice draw upon another, more psychic sense or use their physical senses in a way others of their species and ours have not mastered as yet.

Table 4:
A Sensory Development of Kittens

Visual

Eyes open: 5 to 14 days
Eye color begins changing: 23 days
Depth perception becomes well developed: 28 days
Siamese eyes cross (not present in all Siamese): 6 to 8 weeks
Visual ability at adult level: 8 to 9 weeks

Hearing

External ear canals open: 6 to 14 days
Sound recognition of individuals: 3 to 4 weeks

Smell

Highly developed at birth
Flehmen reaction (lip curl): 6 weeks

Taste

Ability to taste salt: within 24 hours of birth
Ability to taste bitter, sour, and sweet: by 10 days

Touch

Withdraws toe from pinch: 5 weeks after conception
Rooting behavior (pushing of head into warm objects because newborn cannot regulate his own body heat): present until 8 to 10 days
Uses touch for placement of front feet: present by 5 days

Table 5: Locomotor Development of Kittens

Fetal movement: 25 days after conception
Crawling: birth
Walking: 3 weeks
Running: 3½ to 4 weeks
Climbing: 4 to 5 weeks
Air-righting reflex (ability to alter position in midair to land on feet after a fall): 5 to 6 weeks

Intelligence

Intelligence is hard to define, but it is related to an individual's ability to adapt and survive. The cat is intelligent, for he is adaptable to living the life of ease with doting owners or surviving on his own in the wild.

I know a cat that lives with her owners in a mountain cabin located at an elevation over 8,500 feet. She lives outdoors but has access to barns and other outbuildings. Even when the owners spend winters in the lowlands, this cat manages quite well on her own, but is pleased to accept handouts when they are available. Using survival as my criterion of intelligence, she would rate high on a feline IQ test.

Intelligence can also be broken down into components of behavior that are innate or inherited and those that are learned. For the most part, behavior necessary for survival is inherited. Hunting behavior has both components; the movements involved in capturing prey—stalking, pouncing, and grabbing with paws—are inherited, but proficiency traits such as timing and killing are learned.

Interestingly, the higher the level of thinking ability attainable by an animal, the longer it takes him to reach it developmentally. A cat grows up and develops both thinking and reproductive capabilities sooner than a monkey or a person. Yet the monkey and person would receive higher scores on the thinking part of a universal IQ test.

How Cats Learn

The newborn's brain is fully functional at birth and develops rapidly over the first five months of life. By two years of age, the average cat has reached his intelligence peak. Learning, however, is a lifelong process, ceasing only at death.

OBSERVATION AND IMITATION–Cats learn by observation and imitation, just as you and I do. Experiments have shown that cats watching another cat press a bar or jump over a barrier for food will learn this behavior much faster than those without teachers. As discussed in Chapter 2 (pages 25–26), young kittens learn hunting behavior by observing and imitating mother.

Marie, a Persian stray adopted by Dorothy, the delightful lady mentioned in the Preface, is a good example of observational learning. At the time of Marie's arrival into Dorothy's household, Dorothy was teaching the Siamese, Mr. Merry, to ring the doorbell to be let outside. Dorothy didn't bother trying to teach Marie because she figured Persians aren't that bright, compared with Siamese, that is, and surely Marie was incapable of learning such a complicated trick. In spite of Dorothy's training efforts, Mr. Merry refused to ring the doorbell on a routine basis; however, Marie picked up the trick by observing the lesson from afar.

TRIAL AND ERROR–Cats learn by trial and error, as well as by observation and imitation. A kitten bent on getting a shiny Christmas ornament off the tree may try trial and error—first he jumps from the floor; when that fails he climbs onto the bookcase and makes the leap that lands him the ornament, plus the entire tree.

Barbara, our preacher's wife, recounts the example of her cat Flower using trial and error to solve a life-threatening situation. Flower, who had been severely injured, dragged himself home. He tried to attract Barbara's attention by howling outside the front window, but each time Barbara came to the door and looked out she failed to find the cat buried in dense fall leaves.

Finally, Barbara heard a loud crash outside the house. In spite of a fractured shoulder, Flower had managed to attract her attention by pushing over the heavy driftwood sculpture adorning the front of the house. There she found her injured cat.

After she rushed Flower to the vet's office, Barbara was warned that the cat might lose part of his paw or leg. "Not this cat," Barbara said. "We'll be praying for him." Of course, Flower made a complete recovery.

PROBLEM SOLVING–Cats are good problem solvers. If a cat you've put in a cage learns to unlatch the cage lock, be assured that he's going to use that knowledge again to what he perceives as his advantage. Again, I am going to use the example of Dorothy and the Persian Marie to reiterate this point.

After Marie (a male cat) picked up the trick of ringing the doorbell to be let outside, he became an ingenious problem solver. He gained Dorothy's attention by ringing the doorbell; however, instead of waiting by the back door to be let outside, Marie ran to the refrigerator and waited there.

Dorothy, who is also easily taught, learned that she was being summoned to provide waitress instead of valet service. In time, Dorothy installed a dinner bell in the kitchen, and Marie adhered to proper bell etiquette by ringing the appropriate bell for the action she wanted Dorothy to perform.

One study credits cats and monkeys with exhibiting much greater problem-solving ability than dogs. Another experiment set up to determine the differences between dog and cat memory proved that most cats have superior recall when compared with most dogs. This correlates with my experience: Cat patients remembered me, less than fondly in most instances, even after several years.

Naming

Naming, one of the first acts of ownership, gives identity to your pet. I believe it is our responsibility to give our pets individual, appropriate, and uplifting names.

Phonics

Animal researchers suggest that short, two-syllable names are easier for a cat to learn than ones with many intricate syllables. Fluffy would be easier for your pet to learn than Mouse Se Tongue. In addition, cats are reported to associate short, high sounds with happy, pleasant situations and low, deep sounds with unpleasantness and aggression. Sweetie probably sounds more pleasing to a cat than Grumpy; the cat may have a reason to live up to either of those names.

Registered Names

Most registered cats have at least two names, including the name of the cattery where they were born and their personal name. The name of the person or cattery purchasing the cat can be added later if desired. For example, Ociville Chocotah of Ocicountry, a grand champion ocicat born in the Ociville Cattery, is now owned by Jerry Katzky and Don Bozeman, owners of the Ocicountry Cattery.

Personal Names

I have observed that the personal names for pets are often descriptive—Pumpkin and Halloween for an orange cat or Ebony for a black one. We usually think that a Fang or Killer will be just that, while a Baby or Precious will be dependent and cuddly. I can testify to one Killer, a large male tabby determined to banish me from his home and possibly from the living. I elected to diagnose and prescribe for his bladder problem from across the room.

Pets represent different things to different people, and the pet's name may reflect that meaning. Animals that become surrogate children are often given human names. Susie and Garfield are usually perceived by their owners to be imbued with human feelings and emotions. I have a friend who names all his animals and plants Melvin. "Cuts down on confusion," he told me. That was true when Melvin was a cactus, a goldfish, and one cat, but less helpful when the feline Melvin gave birth to five little Melvins. A client changed his calico cat's name from Callie to Florence, his ex-wife's name, after the divorce. "That cat was all I got out of my divorce," he said, "and I get a kick out of yelling, 'Get off the furniture, Florence,' or 'Come here, Florence.' "

The name we select for our cat reveals much about us and our relationship with our pet. Time spent considering the best and most fitting names for our pets is well spent.

Cat Language

Cat language includes purring, vocal sounds, and body language.

Purring

Theories about the cat's unique method of communication, called "purring," are many. Some believe that the purring sound results from the vibration of blood vessels in the chest resonating to the upper airway. The explanation I prefer is the following: Purring results from the stimulation of the muscles of the larynx, or voice box, causing air to vibrate through and around the vocal cords and other parts of the upper respiratory system and creating a noise during inhalation and exhalation of air. This stimulation, conscious or unconscious on the part of the cat, is controlled by a neural message from the cat's brain.

More important than *how* he does it is *why* the cat purrs and what he means by it. Purring during nursing is one of the first communicative exchanges between mother and babies. They seem to be telling each other that all is well. Later, cats communicate among themselves—during grooming and mating—by purring.

Stroking by a human mother-figure often evokes contented purring from the cat. Yet purring is often indicative of feline emotions other than that of contentment or pleasure. Cats may purr when they are nervous, excited, injured, or dying. Therefore, purring on the part of the cat communicates different meanings at different times, depending upon the situation. This is similar to a person who cries in response to fear, pain, sadness, happiness, and merriment, or another who laughs both at funny jokes and when nervous.

Vocal Communication

Feline vocalizations range from 750 to 1520 Hz in pitch, with the lower sounds usually associated with fear or aggression and the higher sounds with pleasant feelings. Communications researchers have determined that cats can make roughly fifteen simple sounds and combine them into ten more complex phrases. Cats say more than "meow," and even that sound can vary from a meek, polite salutation to a loud, aggressive "mee-ow."

I never defined my cat C.K.'s "words," but I seemed to know when he wanted me to "come" (he trained me), when he wanted to go outside, when he was hungry, and when he was mad or fearful. When I transported C.K. in the car, his eyes dilated and he alternated panting with yowling; I didn't follow my own advice of introducing C.K. to car rides when he was an impressionable little fellow of two to three months, and he reacted with alarm to what he considered a steel monster.

Another cat, a nasty patient named Tom, screamed at me like a panther when I approached to administer routine vaccinations. I prayed for that cat's continued good health and tried my best never to approach, at least not without Tom's owner or a chair in front of me for protection. On the opposite end of the patient spectrum was a tiny kitten suffering from constipation that offered such a meek, almost soundless "mew" that one's heart melted in sympathy.

Body Language

Some astute and intuitive owners can interpret a cat's intentions from the slightest quiver of a feline whisker or the flick of a tail, but others totally misread their pets' body language. In the second category are owners who give me more credit than I deserve, as far as winning the good graces of my patients go. For instance, Chirper purrs loudly while sitting on my small portable exam table and begins to swish his tail vigorously back and forth. "Oh, look," the unsuspecting owner says, "Chirper likes you; he's purring and wagging his tail." Of course, Chirper and I are not fooled—we know that purring, in this case, is a sign of apprehension and that the tail language means something like this: I am extremely peeved, and you have sixty seconds or less to cease and desist, or I won't be responsible for the consequences to your person!

INTERPRETATION–It usually takes a combination of cat language for us to understand what our feline charges are communicating. Purring can mean pleasure and friendliness or fear and nervousness. A cat that licks himself may be relaxed and contented or stressed and anxious.

Here's a short test. Suppose you enter your neighbor's garage to borrow the lawn mower and you encounter Fang, a male Rex, standing guard. Fang exhibits the following posture: arched back, tail dropped low and lashing back and forth, hair standing on end, mouth open and teeth bared, and ears laid back. That's right—Fang is protecting his master's lawn mower and what he considers his territory. If I were you, I'd call the kid down the street to do the lawn.

Here's another scenario. When my friend Angie, armed with a brush, approaches her Persian cat Barney for grooming, the cat lowers his body, flattens his ears, and watches her intently through dilated pupils. If Angie continues her approach, the cat runs away and, if cornered, will hiss and attempt to scratch and bite. Barney is fearful when he sees the brush in Angie's hand because the previous owner hurt him once when attempting to remove mats with a slicker brush.

I instructed Angie to show the brush to Barney every day so the cat no longer associates the grooming instrument with pain. She might nonchalantly sidle up to the cat while he is busy eating. If he notices the brush and appears agitated, she puts it away in her pocket for another day. When Barney becomes accustomed to the sight of the brush, Angie cradles the cat in her lap while holding the brush in one hand. Later, she will introduce brief periods of brushing while petting Barney. If the cat cooperates, even briefly, Angie rewards him with a food tidbit and releases him immediately. Hopefully, he will come to associate brushing with two things he wants—food and petting—rather than pain.

Let's try a more positive example. You visit your son's house and find his cat Bozo is sleeping on the den couch. You sit down next to Bozo, and the cat wakes, briefly raises his head to acknowledge you, rolls on his back, and stretches his legs while yawning and gently twitching the tip of his tail. Bozo is saying that you're a pretty good human and offer no threat to him. A suspicious cat would not expose his vulnerable belly to you. If Bozo elects to rub his face and body on you, he will be giving you a more active and enthusiastically friendly greeting.

Of course, every cat is an individual, and you can learn to communicate with your own feline by taking the time to stop, look, and listen when he is communicating with you.

Table 6: Body Language

Rubbing whiskers on object: curiosity

Rubbing face and whiskers on you: love and ownership

Kneading with paws: loving, infantile motion reminiscent of kneading mother's breast

Dilated pupils: fear

Half-closed eyes: relaxation

Pupils constricted: aggression

Staring: warning, "Back off!"

Arched back: an offensive posture meant to intimidate

Ears pricked forward: curiosity

Ears erect and facing forward: alertness

Ears backward: warning

Ears flat against head: preparation to fight in defense

Ears erect and facing backward: declaration of war

Tail straight up: nonchalance

Bristled tail: fear

Relaxed, gently moving tail: contentment

Lashing tail: warning

Tail rigidly behind: aggression

Exposing belly: trust and vulnerability

Sneering expression (drawing back upper lip): smells something interesting or enticing

Quick flick of tongue over lips: agitation

Open-mouth panting: pain or fear or overheated state

Table 7:
Cat Sounds

Formed with mouth closed: grunt, purr, call, sound of recognition

Formed with mouth open; then mouth slowly closes, forming vowel sounds: mild form of mating call. Siamese cry, fussing vocalization, anger wail, expression of puzzlement

Sounds indicating strong emotion: growl, snarl, hiss, intense mating cry, scream

Question

Dear Dr. Whiteley,

I am looking for an adult cat for my mother, who lives alone in a small apartment. I want to select a cat that is friendly, calm, and harmless. What should I look for?

Concerned Daughter in Mountain Home

Dear Concerned,

Closely evaluate each prospective feline companion for social behavior. Schedule a time to be alone with the candidate in a quiet room or other stress-free environment. If it is not his home, give the cat ten to fifteen minutes to become acquainted with the room before you enter and begin your evaluation.

Enter the room quietly and calmly, and stop about six feet away from the cat. Squat and extend one hand forward with the palm up, and call the cat by name. If the cat meows and approaches in a friendly manner to sniff your hand, give him a point. If he retreats or shows any type of defensive posture, subtract a point. Lower your hand and move closer to the cat. If he licks or rubs his body against your hand, he gets another point. If he retreats, strikes out, or bites your hand, subtract appropriate points. Purring and chirruping sounds on the part of the cat gain points; hissing and growling lose points. If you can do so safely, pick up the cat in your arms. If he relaxes and extends his paw to your neck, give him several points. If he struggles or shows fearful or aggressive behavior, subtract points.

You can continue the evaluation by checking his response to noise and toys, etc., until you have made a decision about the cat's social behavior and suitability as your mother's companion.

Best wishes!

H.E.W.

4

Etiquette Training

Potty

In ancient days a mother cat protected herself and her offspring by burying feces, thus hiding their presence from predators and from rodents, their intended prey. This behavior also helped reduce the ingestion of parasite eggs spread through excrement. The mother was unaware of this medical benefit; fate, however, may have smiled on those of the species that began life parasite-free.

Away from home cats prove less prudent about burying feces and, in some cases, leave the fecal deposit exposed on a knoll for other cats to see or smell. It's another way of claiming territory, and males indulge in this type of marking more often than females. I know of one cat family where the male cat uses the litter box but does not bury his feces. His female littermate follows him to the box, sits quietly nearby while he does his business and leaves the box, and then reaches over the side of the box with her paw to cover his feces with litter. Their owner says that the female sibling has been cleaning up after her brother in this way ever since they first went to the litter box together as kittens.

A mother cat maintains the sanitation of her nest by ingesting feces and urine from newborn kittens after they have nursed. The newborn is stimulated to eliminate by the mother's warm, wet tongue rubbing the genitals. When kittens are old enough to follow the mother, usually three to four weeks of age, she leads them to a selected elimination location and licks the kittens' genitals to stimulate them to urinate and defecate in the selected potty. Mother then covers the waste by scratching, and the lesson, whether performed in a rural barn, suburban home, or urban high-rise apartment, is much the same.

A cat mother that serves as a proper nursemaid and role model to her kittens is a jewel we often take for granted. If you are lucky enough to be mentor to a cat mother, provide her with the space and equipment she needs for training the kittens. Offer a litter box with low sides easy for kittens to step over, no more than a couple of inches high, filled with a thin layer, about a quarter of an inch, of clean, dust-free litter and located in an area within easy access of mother's nest.

In the case of orphan kittens, you must mimic mother cat's

behavior. After kittens have been bottle-fed, stimulate them to urinate and defecate by rubbing their genitals with a washcloth or cotton ball soaked in warm water. As soon as the babies are old enough to stand and squat, carry them to a short-sided litter box filled with a thin layer of litter. Rub the genitals with the warm, wet cotton ball once or twice. When the kittens begin to squat and pass waste, express your admiration and pleasure. Cover the waste with your fingertips or a stick and entice the little fellows away from the box. After all, potty training is serious business, and trainees should be discouraged from playing in the litter box. Return the kittens to the box after each meal, nap, and play period, and reward appropriate litter box manners with praise.

Litter and Boxes

Feline preference for litter and litter boxes doesn't always coincide with human preference, and it helps to know what your cat likes and dislikes. Many cats turn up their nose in protest against the dyed and deodorized litter we would choose. You may think that a litter box disguised as a sand castle is the cat's meow, but your pet may not; when given a choice, most cats select open boxes. Litter box cleaning may rank with dusting, a chore that went out with the term "hausfrau," but your fastidious cat may not be so liberated. You may think your cat won't care that his box is located next to your teenager's room, but your cat may prefer more privacy and less hard rock than your teenager. In other words, cats have strong opinions about litter boxes and litter products, and if your cat companion is reluctant to use what is being offered, you might design your own feline preference test by proffering different types of boxes and litter box fillers.

Cats are animals that produce concentrated urine, which, when exposed to bacteria and air, causes odor, and litters work to control that odor in three ways—by absorbing moisture, reducing its exposure to bacteria and air; by adding a deodorant to mask the odor; and by using additives such as antimicrobial agents to reduce production of odor or to neutralize it. Most commercial litter products utilize one or all of these methods.

Litter or catbox fillers have been made from sand, sawdust,

newspaper, wood shavings, and just about everything else the human mind can think of, but most commercial varieties use some form of clay. In a preference study conducted by Dr. Peter Borchelt, an animal behavior consultant, most cats selected a clumping-type litter, and few of the cats in the study liked wood- or newspaper-type litters.

Although the earliest variety has been around for several years, litter designed to absorb liquid waste into clumps, which can then be flushed or disposed of in the trash, is experiencing a surge in popularity. Several brands and formulas are available, the main difference being the size and compactness of the clumps that are formed. For cats that do a lot of scratching and digging, the clumps must be larger and harder. To test the workability of clumping litters, bounce a soiled clump on your litter scoop. If the litter remains intact, it is doing what it's supposed to. Clumping litters are not designed to be disposed of through septic systems, and I am personally reluctant to flush them through my older house's cranky plumbing.

Another new product is a home-test litter for the detection of diseases diagnosed by urinary changes. The test litter is added to the top of your cat's regular litter; a color change in particles of test litter indicates a problem. These tests, dispensed by your pet's veterinarian, are screening tests only, and positive results should be confirmed in the veterinary laboratory. At-home urinary tests are a helpful aid to owners who must monitor cats with chronic diseases such as diabetes and feline urinary disease.

Table 8: Litter Box Suggestions

Adjust the location of box, height of box sides, and depth of litter for the convenience of kittens, elderly, and ill cats. Place a towel or rug under the litter box so the cat can wipe his feet.

Long-haired cats may prefer a coarse-grained litter, rather than a fine-grained variety, because it clings less to fur. You might consider setting up your own test to determine your cat's likes and dislikes.

Place litter boxes in locations that are easily accessible, reasonably secluded, and quiet.

For most adult cats, place 2 to 3 inches of litter in box.

Offer one box per cat when possible.

Remove both urine and feces from boxes and add fresh litter daily.

Completely change litter when necessary. Let your nose and eyes be the judge.

Clean the box with a mild detergent; never use a strong-smelling disinfectant. Avoid ammonia products because they smell like urine.

Do not disturb your cat while he is in the box.

Clean up accidents as soon as possible. Use one of the newer products made specifically to break up the source of urine odors such as Cat-off, F.O.N., and Outright. Rug cleaners are also helpful.

Commode

I admit that I've never been caretaker to a cat that used the commode; however, I do know Percy, the commode-using office cat at Village Animal Hospital.

Village Animal Hospital is an all-female establishment, and Percy is also female, although I suppose it doesn't matter. Percy, anyway, taught herself to use the commode by watching the ladies in the hospital. Everyone in the office was convinced that someone was careless about flushing, but no one wanted to broach the subject. One day the culprit was discovered in the act of urinating in the commode. Percy certainly proved that cats learn by observation.

Not every cat becomes potty-trained as easily as Percy, and there are books devoted to teaching your cat to use the toilet and commercial kits available to make the training easier. My friend Willie at the local pet store sells a kit consisting of a plastic form that fits between the toilet seat and porcelain rim; inside the form is a compartment for litter containing holes that allow urine to drain through. A packet of catnip to make the litter more attractive to the cat is supplied (not all cats are attracted to catnip). If you are creative, you can make your own version.

It appears to me that there are two ways to approach commode training: Make the commode into a litter box or make the litter box into a commode. The former requires that you give up one of your toilets until training is complete, and the latter requires the acquisition of an additional toilet seat.

Training will go more smoothly if your cat is already using a litter box located in the bathroom. You can add the toilet seat to the litter box and gradually increase the height of the sides of the box until your cat is used to standing on the toilet seat and urinating and defecating into the litter below. Litter can eventually be replaced by water. Hopefully, the cat will adjust to the regular toilet after he has been trained to use the makeshift one. If you make the toilet into a litter box, you gradually reduce the litter contained in the box portion until the cat becomes used to urinating and defecating directly into the water in the toilet bowl. If your training goes too rapidly and the cat backslides, so to speak, or if the cat actually falls into the toilet, you'll probably have to start over with your training.

It's efficient, I suppose, to have all household members, including the cat, use the expensive Kohler plumbing fixtures. I would prefer, however, a cat that could take the toilet brush, Sani-Flush, and Ajax and clean those bathroom fixtures. Now, that would be a real potty-training accomplishment.

Outside

I have never been able to figure out why a cat is always attracted to a neighbor's flower bed, rather than his own, to use the potty. It may be that most outside cats don't show this propensity, but those that do stimulate vocal complaints from neighbors whose flower beds and yards are chosen. It may also be that the neighbor's beds have been recently groomed, making them attractive places for the cat to do his business.

When cats eliminate, they scratch loose dirt until they make a shallow depression. They eliminate in the depression and then cover it back over with dirt or leaves. A nice flower bed probably has great appeal for a cat. If your cat is creating ill feelings by using your neighbor's yard instead of his own, the remedy consists of making your yard more attractive and your neighbor's yard less attractive. You can get real inventive with the ways you accomplish the task—a catnip-enhanced sandbox in your yard, a water hose sprinkling your neighbor's yard, for example. Of course, observing cat leash laws will also take care of the problem.

Grooming

Claws

Cats are territorial creatures, and that territory can be a room, a piece of furniture, or an outside tree. It is not uncommon to find such a tree, replete with carvings, which attracts the same cat back again and again to claim ownership, not only by the wood markings, but also by scent from the sweat glands located in the bottom of his footpads. The tree also serves as a means of conditioning the cat's claws.

Feline claws are not sharpened like a pencil. The cat scratches to loosen the frayed and worn outer claws that are peeled off, exposing a new, sharp claw underneath. The sharpening can be accomplished with a scratching post, as well as an outside tree, and the sharp nails of the indoor pet can be blunted to reduce the danger or nuisance of these fine-pointed daggers.

PEDICURE–Most indoor pets need regular nail clipping to prevent the pet's overgrown nails from snagging your expensive nylons or catching in the carpeting. For most cats, trimming every two to four weeks is sufficient to take the scratch out of the cat. I prefer to use nail trimmers made specifically for pets, but an ordinary toenail clipper sold at drugstores can be used. When using a human nail clipper, turn it sideways to the claw. Make sure that the clipper has a sharp blade that will cut cleanly, rather than shred the nail.

Before cutting, make sure your cat is calm and relaxed. I like to cradle the cat in my lap while actively engaging my feline companion in what I hope is pleasant conversation. I then grasp the cat's paw with my left hand and expose the claw while I clip with my right hand (I am right-handed). Cats that have been conditioned at an early age to having their paws handled will come to associate nail clipping with extra time and attention from the person performing the pedicure.

If you are starting with a kitten or new adult cat, spend brief periods, no more than a minute or two, letting him examine and sniff the nail clippers. Then progress to handling the cat's paws. You can make a game of it by putting the cat on the opposite side of the door and enticing him to respond with his paws to your wiggling fingers, while you, in turn, touch and grasp the cat's paws. If he becomes agitated, stop at once. The idea is to get the cat used to thinking that brief touching and handling of his paws is play. Hopefully, you and your cat will graduate to your holding the cat in your lap, grasping each paw, and exposing the nails.

For the procedure itself, gently protrude each nail to be trimmed by pressing on the footpad near the base of the nail. If the pet has light-colored nails, clip just below the pink-colored quick, which contains nerves and blood vessels. If the pet has pigmented nails, clip small pieces of nail at a time. If bleeding occurs, hold a small piece of gauze or cotton over the nail until the bleeding stops. If the nails are rough following trimming, use a file to smooth them. It is better to clip one nail per session to begin with than risk having your cat become frightened of you, the nail clippers, and the pedicure.

Pedicare is especially important for the senior pet. As an animal reaches old age, his nails become soft and brittle, contributing to cracking and breaking.

Soothing foot baths followed by careful drying with a soft towel are indicated for animals that tear a nail or suffer skin irritation around a nail.

SCRATCHING POST—For some cats, keeping the claws well trimmed will take care of scratching problems. For others, a scratching post will provide an alternative to the leather couch, mahogany desk, and custom drapes.

Just as cats have preferences for the type of litter you offer, they also have preferences about the texture and form of scratching posts. Your cat may or may not be excited by the scratching post on sale at the discount store or the one covered with carpet that matches your decorating scheme.

Most cats prefer a stringy fabric with longitudinal threads that can provide a nice drag when pulling down on it; the more worn and straggly the better. Often caretakers recover or buy a new post just when the cat has the fabric shredded to his liking.

The post can be attached to the wall or can stand alone, upright or horizontal in orientation, as long as it won't tip over when the cat uses it. For cats that like wood, a rough-barked log anchored to the floor or wall might be just the thing, and for owners who have room for it, two posts are better than one.

A kitten or cat should be directed to a scratching post before he develops bad habits. The post must be prominent and kept in an area the animal frequents. Cats stretch upon waking and also when clawing, so it might be a good idea to locate one post near the cat's bed. For some cats, rubbing the new post with catnip is an incentive to use it. For others, training is required. Rub the cat's paws on the new post so secretions from the sweat glands will mark the post as the cat's very own. You might entice a kitten or cat to chase a string around the post, and, of course, praise the cat lavishly each time he uses the post for scratching.

DECLAWING AND ALTERNATIVES—I received my greatest volume of mail in response to a column I wrote in which cat declawing was mentioned. Yes, most letter writers were against the procedure, but I stick to my premise that declawing is a viable and permanent option for a destructive cat that does not respond to behavior modification. If a cat uses a scratching post and is not destructive, the surgery is needless and should not be performed.

To support declawing as an option, a recent study conducted by Dr. Katherine Houpt, an animal behavior expert at the New York State College of Veterinary Medicine at Cornell, shows that declawing is not related to future behavioral problems such as house-soiling or destructiveness in cats. Owners of purebred cats that might enter the show ring, however, should use other alternatives because most cat shows have rules prohibiting the exhibition of declawed cats.

Declawing surgery is performed under general anesthesia, and the nail, along with the portion of bone containing the nail growth plate, is removed. There are several different procedures, and each veterinarian has his or her favorite method. I suggest that the nails of the front feet only, usually the most destructive, be removed. Cats are then able to climb rough-barked trees, aided by the claws on the back feet, if the need should arise. I am reluctant to declaw outside cats because they are then at a disadvantage if encountering a dog that might be discouraged by a swipe with a talon-exposed front paw.

Another surgical procedure, not used widely at present, involves cutting the tendon of the muscle the cat uses to flex his nails to scratch. This procedure must also be performed with the cat under general anesthesia, and the nails remain intact and must be trimmed regularly.

New to the pet-scratching market is a product consisting of plastic shields that are glued over the pet's nails, similar to false fingernails. The nail caps, called Soft Paws, were developed by veterinarian Dr. Toby Wexler and designed as an alternative to declawing; the covers will not interfere with normal extension and retraction of the nails, come in several sizes to accommodate growing and different-sized pets, and are recommended for use on front or back claws. They can be purchased from a veterinarian, who will apply them in the office or dispense them for you to apply at home. The drawback, of course, is that new shields must be applied each time the cat sheds a nail, usually every four to six weeks.

At first the cat may resist the idea of his false toenails and chew one or all off in short order. In that case, you must reapply the nails. Remove all caps that have been torn by peeling them off with your thumbnail or clipping them off with the nail trimmer. Clean the cat's nails with alcohol, let dry, clip each nail's sharp tip with the nail trimmer, fill the plastic nail cap approximately one-third full with adhesive, extend the cat's nail and slide the plastic cap over the nail until the tip fits snugly over the end of the nail, and hold the cat still for about five minutes to ensure that the adhesive has bonded.

Several years ago a similar product was advertised. Small wooden beads are glued over the tip of the cat's nails. The principle is the same as for the plastic shields and the drawback is the same —the beads are lost every time a cat sheds a nail. Regardless of whether these products are the answer to your cat's scratching problems, the most important consideration is safety to pet and family members. The glue and vinyl shields used in Soft Paws are advertised as safe and nontoxic if swallowed, giving it a definite edge, in my thinking, over the beads.

Hair

Hair has been a source of beauty and adornment for both people and pets throughout history. Proper hair care not only enhances beauty, it also contributes to the comfort and health of your cat.

Cats have three types of hairs: tactile hairs (whiskers), coarse or guard hairs, and fine or secondary hairs. The skin surrounding whiskers is rich in nerve endings, making whiskers into finely tuned sensing devices. Whiskers serve as feelers for a cat, providing another way for him to assess his environment by using his whiskers the way a blind person uses a cane. I have seen young cats in winter with whiskers burned off that must have given in to curiosity by touching the hot stove with their feelers.

The chief functions of primary and secondary hairs are protection and insulation against heat and cold. These hairs do not grow continuously, but are lost and renewed periodically. The growth cycle consists of an active phase, a transition phase, and a resting phase. In humans, at any given point in time, 80 to 90 percent of hairs are in the active growth phase, 10 to 20 percent in resting, and one percent in transition. Detailed studies are not available for animals, but I assume findings would be similar. Hair grows at a rate of a tenth to half of an inch per week, and seasonal shedding occurs in response to changes in daylight hours and, to a lesser degree, changes in outside temperatures. Hair is naturally thinner between the ears and eyes, and this natural baldness becomes more noticeable during shedding.

Cats are blessed with hairs that stand more erect in response to emotional states such as fear and anger. These hairs are more highly developed over the back, rump, and top line of the tail. Remember, the cat with the arched back and bushy tail is a scared or mad cat.

BRUSHING AND COMBING–A routine of brushing and combing removes loose and dead hair; deters natural grooming that can lead to hair balls; removes burrs, stickers, fleas, and flea dirt; and allows you to assess the state of your pet's hair and skin, as well as provides that sense of touch and caring that is vital between caretaker and charge.

The mother cat interacted in a pleasurable way with kittens by licking and grooming them. When you adopt a new kitten, you want the act of brushing and combing to be an extension of a mother's loving touch.

Cats like ritual, and it might be helpful to set up a ritual around brushing and combing by using the same place, tools, and time of day, and by grooming the cat's body parts in the same order. The frequency will depend upon the individual cat's needs, although grooming with your fingers might be a daily procedure for either long- or short-haired cats.

For short-haired cats, a small wire-bristled brush called a slicker brush may be the only tool you need. Start by lightly brushing the back of the neck and under the throat and chest, places a cat usually likes to have brushed. Proceed over the rest of the body, using care to avoid irritating the thin skin of the abdomen, inner thighs, and nipples. As always, you want to use short grooming sessions at first. If the cat becomes apprehensive, put away your tools for now. As your cat cooperates, offer praise and a treat.

Grooming long-haired cats is an art, one with which I personally have little experience because all of my cats have been short-haired. I would like to make the following disclaimer: If you need help, please consult a professional groomer, one who specializes in cats.

In spite of my inexperience, I offer the following suggestions. Start with a wide-toothed comb and progress to a fine-toothed comb. Use short strokes of the comb and allow the comb to become an extension of your fingers, enabling you to detect small mats before they become big ones. Never use water on a mat because water sets the hair forming the mat into a more dense mass.

It is best, of course, to catch mats small and soft, and comb them out with a coarse comb before they get to be large, compact mats. Larger mats can be divided into smaller parts by using the blade of open scissors, and then cut and combed out. Caution is urged because it is very easy to cut the cat's skin underlying the mat. Mats that cannot be combed out may be removed with electric clippers.

With scissors, clip away long hair under the tail that traps fecal material.

Cats have the thinnest skin of any domestic animal, and if your cat is hurt with rough handling or scratched or cut with comb or scissors, it is going to be hard indeed to convince him that hair combing or brushing is pleasant business. By starting early in the pet's life and using gentle grooming practices, you will hopefully avoid circumstances that lead to pain for you and your cat.

Bathing

It is helpful to introduce your kitten to water and bathing when he is an impressionable little fellow of two to four months of age. Remember how Percy trained herself to use the commode by watching her human family. Certain cats will adjust to being placed in a few inches of warm water in the bathtub by watching you enjoying yourself there, while others will need coaxing. Perhaps a rubber ducky and a few frothy bubbles are just the thing to convince a juvenile kitty that the bath is a fun place to be. A warning comes to mind: When you train an animal to be unafraid of something like a bathtub of water, you must be cautious that the bathtub of water is safe. I have treated animals that jumped into a tub of scalding water.

Frequency of bathing depends on the need and condition of the cat's hair coat. Certain fortunate felines may never need to be bathed, while others with skin disease may require weekly bathing.

Make your cat or kitten more secure when bathing him in the bathtub or sink by placing a towel or mat in the bottom to prevent slipping. Protect the cat's eyes by placing a drop or two of mineral oil or ophthalmic ointment in each eye before shampooing. Shield the genitals with petroleum jelly, and place a piece of small cotton ball in each ear to prevent water from getting in the pet's ear canals (don't forget to remove it). Wet the cat thoroughly with warm water, lather, using the proper shampoo, and rinse well.

Select shampoo, cream rinse, and conditioners made for cats, as most human products have a different pH from that needed for animal hair. Cream rinse is applied to wet hair to make it easier to comb; most coat conditioners are applied after the hair is dried with a towel and/or hair dryer. Dry shampoos and foams are available for use when it is not convenient to give a wet bath. Baking soda can be rubbed lightly into the pet's coat and brushed out for in-between baths. If you use a dip for flea and tick control, make sure it is a cat product; mix and use according to directions. In fact, use all products as they are intended.

Teeth/Gums

Compared to dogs, cats have smaller mouths and fewer chewing teeth—two fewer baby premolars and twelve fewer permanent premolars and molars. The early cat, definitely a carnivore, caught and punctured the flesh of his prey, depending primarily upon his canine teeth.

A kitten begins cutting baby teeth at two to four weeks and loses these temporary teeth to the permanent ones beginning at three and a half to four months. Institute rules of good dental hygiene before the kitten loses the baby teeth.

Sweet treats should never be offered. Strong, hard-rubber chew toys given for five to thirty minutes twice weekly help to strengthen tooth attachments. A dry food formulated for kittens is preferable to canned for keeping the teeth clean. Brush your pet's teeth daily (dental plaque mineralizes into calculus, which causes gum disease in twenty-four to forty-eight hours).

Your kitten can be trained early to tolerate tooth brushing. Retract the kitty's lips with one hand, and brush in a circular pattern and in strokes horizontal to the gum margin. You can use your finger or a baby's toothbrush soaked in water, adding toothpaste formulated for cats when the pet gets used to the procedure. Initially, clean only one or two teeth or massage the gums only. Then praise and pet your kitty.

In the last few years numerous products for brushing pets' teeth have appeared, and I'm sure that several more will be available in the future. Some of these include malt-flavored pet toothpaste, mouthwash for pets dispensed in a spray bottle, gauze pads impregnated with dental polish to be wrapped around a Q-tip or your finger and used to brush the pet's teeth and gums, toothbrushes designed for pets, and a rubber toothbrush that slips over your finger and can be used to brush pets' teeth and massage their gums. It is best to use what fits most easily into your cat's mouth and proves to be most comfortable for you and your pet.

Eyes/Ears

Other grooming care includes wiping the eye area daily with a moist cotton ball for all cats that have ocular discharge and carefully clipping with scissors any hair that traps discharge under the eyes. I recommend cleaning the inside of your cat's ears every couple of weeks with a cotton ball moistened with diluted hydrogen peroxide. A cotton-tipped swab may be used to clean the folds in the skin of the outer ear. Refrain from using cotton swabs in the deeper parts of the ear canal, because overzealous use can cause debris and wax to be packed against the eardrum, resulting in damage to the drum or contributing to infection.

Ear care products are available commercially for ear cleaning, dissolving wax, treating ear mites, and drying the ear canals. However, it is best to check with your veterinarian before applying any product inside the ear canals of your cat.

If your veterinarian prescribes medication, usually drops or ointment, to be put into the ear canal, the following procedure can be used: Cradle the cat comfortably in your lap, reach over and hold the tip of the cat's ear facing away from you, gently squeeze two or three drops of medication into the opening of the ear canal with your opposite hand, and massage the ear canal (located at the base of the ear) until a squishy sound is heard with each massage. Talk soothingly to your kitty during the procedure and reward him when it is finished. If your cat is anxious after medicating one ear, let him down and come back later to treat the other ear.

Administering Oral Medication

Tablet/Capsule

Some cats take pills or other medication hidden in a small amount of palatable food such as a bite of cheese or a meatball. In my experience, however, most cats are not that accommodating. Some take the treat and spit out the pill or refuse the treat entirely.

Just in case your kitty is the stubborn type, I suggest administering the oral medication in a neutral area of the cat's territory, not in a place such as his litter box, which you want to remain attractive to the cat. The less restraint or wrestling with the cat, the better; it might be beneficial to have a partner hold the cat while you do the deed. Talk softly and soothingly rather than yell at the cat for behavior that makes sense to him. Praise any sign of cooperation on the part of the kitty.

Medications can be administered with the cat standing, sitting, or lying on his chest on the floor, on a bed, or on a small table.

Wash your hands, lubricate the capsule or tablet with a minute amount of butter, hold the capsule or tablet between the thumb and index finger of one hand, place the fingers of your other hand over the top of the cat's snout, rolling his upper lip inward over his teeth (if he bites down he'll bite his upper lip), raise the cat's head upward with the hand holding his upper jaw, press down on the cat's lower jaw with one or more of the remaining fingers of the hand holding the medication, place the capsule or tablet on the base of the cat's tongue as far back as possible, withdraw your hand quickly, shut the cat's mouth, and stroke the neck area until he swallows.

There are some refinements to this method. Those that feel trusting or brave might push the pill down the hatch with the index finger while that vulnerable finger is in the cat's mouth. Commercial pill pushers are available at veterinary hospitals and pet stores; these handy instruments work on the syringe principle—a plunger pokes the pill down. Some people have better luck with the pill or capsule unlubricated, as too much butter can cause the pill to stick to one's finger instead of going down the hatch.

The outcome of medicating cats is not always the expected one. An owner may congratulate himself for his accomplishment, only to find the pill lying on the carpet when vacuuming the next day. Another may have the ungrateful cat spit the pill back into his face. On rare occasions, one may become so intent on giving the medication that he fails to notice that the cat is choking or having difficulty breathing. Knowing when to stop is also part of the art of nursing.

Liquid

Although some cats will take liquid medication from a spoon, I find it easier to administer this form of medicine in a dropper or small syringe.* The cat's jaws can remain closed for this procedure. Make a pouch in the corner of the cat's mouth by placing a finger or thumb inside the cat's cheek and pulling out on his lip. The cat's head should be parallel to the ground or slightly raised. Place liquid medication into the cheek pouch, small amounts at a time. Proceed slowly, waiting until the cat swallows before adding more medication.

Pastes

Laxatives and nutritional supplements may come in a paste or gel form. These medications can be administered to the side of the cat's lips or even to his paws. Most cats will lick the medication off their face or fur.

Question

Dear Dr. Whitely,

Can CPR be performed on cats? My neighbor's big tomcat was hit by an automobile in the street in front of our house. The cat ran under the bushes in our yard, and when I found him he had just died. I wonder if I might have saved him by administering CPR.

Cat Lover in Arlington

Dear Cat Lover,

I don't know if the neighbor's cat could have been saved with cardiopulmonary resuscitation (CPR), but it certainly would have been worth a try.

First, assess whether or not the cat is breathing. You can do this by observing his chest rising and falling and hearing and/or feeling the air going into and out of his nose or mouth.

* Syringes and droppers are usually calibrated in milliliters (a milliliter is the same as a cubic centimeter). To convert teaspoons to milliliters, 1 teaspoon = 5 milliliters.

If he is not breathing, lay the cat on his right side and extend his head. Use your hands to seal his mouth and to form a tube through which to blow air into his nostrils. Move your mouth away from your hands when allowing the animal to exhale. Give the cat a quick breath every five seconds. After four or five ventilations, check to see if the cat is breathing on his own.

If he is not breathing, check to see if his heart is beating. Feel and listen over the lower third of his rib cage just behind his elbow. If there is no heartbeat, you or a partner must begin external heart compressions while continuing artificial respiration. You may need to move the animal onto a firm surface and to stabilize his back to keep him from sliding away from you.

Place the heel of your right hand (if you are right-handed) over the lower third of the left chest at the level of the fourth to sixth rib. Exert moderate and smooth pressure compressions at a rate of one to two per second. After five to ten compressions, give a breath. Continue at a rate of twelve breaths and 60 to 120 heart compressions per minute. Stop periodically to see if the cat's natural heartbeat and breathing have returned.

Good luck!

H.E.W.

5

Country Cat/City Cat

Hunters

Cats are marvelous hunters, moving with the fluid grace of ballet dancers from crouching position to high-speed running, from stalking to pouncing. The feline back is flexible, when compared to that of a forager like the cow, enabling the cat to catch prey by adding force and distance to his stride. Agility is an attribute of the cat's slim shape. Like a warrior, the hunter must be a mean, lean fighting machine.

Because the cat has survived and evolved as a hunter, nature has equipped this soldier for the task. The slight body shape is possible because the cat has a short digestive tract for rapid digestion. The feline predator does not carry around excessive food bulk like a grazing animal. Feline babies are born at a relatively immature stage, thus freeing the mother of the physical handicaps of pregnancy. Mother and babies alike fare better if mother is free to hunt and bring back food for her helpless brood.

The cat's sharpened, retractable front claws are useful daggers, and feline teeth are perfectly designed for killing small critters. When a cat makes a killing bite to the nape of a mouse, for instance, the cat's sharp-pointed canine teeth, flattened on the sides, permit the clean severing of the victim's spinal cord without damaging its bony spine. The bite is often made so cleanly and efficiently that it is difficult to find on casual observation. The cat eats the quarry from the side of his mouth, using the scissorlike action of premolars.

Cats that live as hunters are usually successful. Most cats prefer rodents first, then rabbits and other small mammals, and birds; some cats become specialists at catching a particular type of prey, and specialist mothers teach those skills to their kittens. The fishing cat of Malaysia, for example, is skilled at catching fish with his front paws.

Animals that bring down large prey usually hunt in packs; since the domestic cat is a solitary hunter, he is restricted to animals smaller than himself. As mentioned, cats are smart, choosing not only smaller quarry but quarry that is more defenseless than they. For this reason, few cats will tackle an adult Norway rat, an animal that fights back viciously when attacked.

Cats are patient, often waiting an hour for a mouse to leave its den. The feline then stalks the unsuspecting mouse, slinking after it and waiting for the opportunity to pounce or slap it down with a paw. As the cat springs forward for the catch, his whiskers protrude as far forward as possible and aid him in checking the prey's movements after capture. A small mouse is practically encircled by whiskers. After the catch, the cat puts one paw on the prey and grasps it. Some cats will play with the mouse before killing it with a penetrating bite to the neck, while others, those that are well fed, probably, let the mouse get away.

The technique is modified for bird hunting. The cat waits, crouched behind a bush or frozen in stealthy pose, body low to the ground, one foot raised, waiting for the bird to dart close by so he can pounce on it. If the bird takes to the air before the pounce, the cat raises himself on his hind legs as the bird lifts off, bending his paw and swatting with lethal claws like a frantic butterfly collector attempting to net his quarry. The hunt is over quickly with the bird flying away out of reach, more often than not.

A friend whose cat Pepper is a legendary hunter reported that Pepper was in her backyard sunning when a nasty bluejay dive-bombed her. Pepper closed her eyes to half-slits and stretched out to expose her belly, a posture of vulnerability, but when the bird approached again Pepper changed from a seemingly harmless pussycat to a ferocious combatant, catching her tormentor and delivering it like a trophy to my friend. "Served that bird right," my friend said.

"Do you think Pepper used rational thinking and baited that bird by faking her subordinate behavior?" I asked. "Absolutely," my friend replied proudly, "Pepper is smarter than a bluejay."

Encouraging Hunting

What do you do to encourage hunting? As discussed earlier, you select a kitten from a mother with good hunting skills and leave him with the natural teacher long enough to learn the behavior. If you have a rodent problem in one particular location in the house or garage, you can introduce your cat to that area of the house and entice him to stay there. Although a fat, lethargic cat is obviously a poor hunter, I don't recommend starving a cat into hunting. I would prefer to feed him in the mouse-infested area so he considers it his home, to be protected from rodents.

Discouraging Hunting

How do you discourage hunting? Declawing will reduce the hunter's effectiveness, but I hesitate to recommend that procedure for outside cats since it leaves them vulnerable to neighborhood dogs.

One study suggests that male cats have a lower threshold than females for hunting behavior; therefore, castration of toms may decrease their desire for both girlfriends and prey. Perhaps, the male hormone testosterone plays a role in the thrill of the hunt. Most females, cat or human, are too smart to sit in duck blinds at 4 A.M. in a wind chill of minus 20 degrees Fahrenheit in an attempt to capture a feathered creature smart enough to be heading south toward warmer weather.

A bell to be worn on the cat's collar or other warning device is the best approach to reduce successful hunting in the outside cat. The indoor cat that stalks the family canary or goldfish with malice must be discouraged in other ways. Making sure the bird or fish is safely housed in cat-proof accommodations and administering remote punishment such as squirting the cat with a water pistol when he approaches the canary's cage or fish's bowl would be one method. Making the canary or fish's room off-limits by booby-trapping it with loaded mousetraps would be another. The best method is socializing the young kitten to birds and fish at an early age or retraining an adult cat to tolerate feathered or finned animals. Training methods will be discussed in Chapter 7.

The Social Scene

Feral Cats

The domestic cat is unique in that he can live with or without man. He may live as a wild creature, depending upon his own resources for survival. Feral cats, those that have reverted from domesticated to wild status, have been found, for example, in remote regions of the world.

One study of feral cats living on an island and feeding on rabbits reports that the cats became so adept at hunting rabbits that they actually ate themselves out of house and home. When the rabbits were killed off, the cats had nothing to hunt and they, too, died.

City cats rummaging through garbage dumps and farm cats receiving minimal subsistence from the farmer are next in independence. These free-roaming cats may live as solitary individuals or as members of a social group.

The Sisterhood

A social female group consisting of a female matriarch and several generations of her female offspring often reside on farms. Cats that are biologically related smell more alike than those not related, and for this reason a kitten from a daughter's litter, for instance, will be more readily accepted for communal rearing and nursing than a kitten from an unknown litter. The cooperating members of the sisterhood also hunt together and act collectively to repel outsiders.

The sisterhood may accept a strange male into the group at mating time or rebuff him if they are protecting a nest of kittens, but female outsiders are verboten. For the most part, young male offspring of the sisterhood leave home to seek their fortune as soon as possible, while female kittens stay at home.

The Brotherhood

During breeding season, free-roaming male cats may form a loose brotherhood. The brotherhood is led by a don, the dominant tom that tries to monopolize the matings in his home range. He is the tattered male with cauliflower ears, scarred neck, and huge jowls.

The don may or may not tolerate a group of two- to three-year-old male hangers-on that occasionally challenge his superiority. Outcasts are subordinate males that avoid contact with the brothers, as they are chased away if their presence is noticed.

Occasionally, a community social of both free-roaming males and females will be called in a neutral area during early evening. The cats will sit in a loose circle about five yards apart, socializing congenially for a couple of hours before dispersing to return to their individual homes.

Social Rank and Territory

Territory is defined as that area of space which an animal considers his own and which he will defend against members of its kind. Robert Ardrey, author of *The Territorial Imperative,* writes: "The disposition to possess a territory is innate. The command to defend it is likewise innate. But its position and borders will be learned. And if one shares it with a mate or a group, one learns likewise whom to tolerate, whom to expel."

Ardrey believes that animals, including man, are instinctively territorial, and he makes the convincing point about human territoriality that more lives have been lost in wars devoted to protecting the motherland than for the love of a woman. David E. Davis, quoted in *The Territorial Imperative,* compares the aggressiveness of teenage street gangs with that of animals defending territory. The goals are rank and territory, he writes, the same as those for which animals fight. Insight is gained by knowing what we fight for, regardless of whether "we" are human or feline.

Cats living in groups—free-roaming cats and pet cats sharing a house—tend to rank themselves according to dominant traits. Right to territory is one of these traits. If a new cat is introduced to a resident cat, the resident cat is dominant by nature of his claim to his territory until proven otherwise.

Intolerance of Resident Cats to Strange Cats

A stranger in their midst sparks discord within a group of cats that normally live together congenially. Not only is the new cat inspected and possibly attacked, but familiar cats also scrutinize and sniff each other as if they were strangers.

I noticed this same kind of intolerance when I transported two sibling American shorthairs into the clinic for spaying. I asked the owner if the two were compatible, and when informed that they were I put them together in the large transport cage. One meowed apprehensively during the trip to the clinic, but they both seemed glad the other was along for the trip. However, after surgery a day later, when I took the cats out of their respective hospital cages and placed them in the transport cage for the trip home, the two hissed, swatted, and screamed at each other. This situation occurred nearly every time I tried to place two housemates together for the ride home, and I finally learned to provide each cat with his own cage.

At the time I attributed the intolerance to the stress of surgery. After all, I'm more cranky with family members when I'm scared and sore. However, it may be that the strange territory of the hospital was the inciting factor, or more likely, it was a combination of things that made two familiar cats strangers to each other.

I also observed that cats remaining at home treated the returning cats as newcomers, often challenging them aggressively. I chalked it up to hospital smell, but perhaps territoriality incited the dispute.

Life goes on when one is away, and the popular cheerleader or football hero who moves elsewhere and returns must prove herself or himself again; you just don't have to be gone as long if you are a cat.

DOMINANCE–It seems that one cannot get away from the Social Register even in cat households, although cats do not display a pecking order like chickens, in which one chicken picks on the next chicken lower in rank, and so on, down to the most inferior. The hierarchy is arranged more loosely into a top cat and lesser cats, and feline interactions within the household may be harmonious or discordant, depending largely on the nature of the dominant cat.

This cat may compete as top cat by fighting others over food, toys, and resting places or may foster a first-come, first-serve attitude. Once a cat establishes himself as the dominant cat, the others may challenge his authority but rarely fight each other for a secondary position.

A dominant tom occasionally becomes a despot, walking around with raised hair and stiff back. This cat displays his dominance by seizing other cats, male or female, and mounting each, not for sex but for a show of his power.

A cat also expresses dominance by adopting high places. A cat that can pounce from above has an advantage, although few actually jump directly onto another cat. Most will land beside the other cat before attacking. Cats dislike being looked at directly, and a dominant cat may exert his tough guy image by staring at a competing cat until that cat backs down by looking away.

RANK AND TERRITORIAL DISPUTES–In spite of the image of the dominant cat as a male bully, the dominant cat is not necessarily the largest cat and may be female. Dominant females fight to protect their home, and territorial disputes occur more frequently with females than with toms. Toms engage in battles over rank and sexual favors. Neutered males tend to fight like females for territory.

I am tempted to compare these feline tendencies with women who will not share their kitchen or recipes with others, and men who compete for status and women and boast of their conquests. The conclusion that human behavior mimics animal behavior rather than the other way around is shared by the previously cited writer who studied street gangs.

Cats in a multiple-cat household may elect to avoid each other's territory and come together only occasionally: to socialize with owners and to eat. Most disputes between cats occur in areas of social and feeding interchange where it is difficult to practice evasion.

In one client's home there were two distinct cat households. The original female cat occupied the top floor of a two-story townhouse, and the rest of the cats, varying from five to nine at a time, lived downstairs, but never did the two separate households meet. The staircase was the invisible dividing line, and it proved as effective as a solid door.

One of my favorite true stories of pet heroism is that of Renee, a young career woman who lived alone in a large house with her Siamese cat, Simon. Simon was very territorial about the house and protective of Renee, and challenged anyone who climbed the stairs to the upper level where the two slept. One night Renee awakened to Simon's shrill howling and the sound of scuffling and fighting. "The first thing that crossed my mind was that another cat had somehow gotten in the house and Simon was involved in a cat fight," Renee said. She started down the stairs to investigate, and saw that the television and VCR were missing from the living room; she entered the kitchen in time to see a man run out the outside door with Simon close on his heels.

Ten minutes later Simon returned, uninjured but covered with human blood. The police credited Simon with saving Renee and her belongings from the burglar. Even without claws, Simon was one tough kitty that must, in this instance, be praised for protecting his territory.

Table 9:
Territorial Marking

Visual: scratching of trees, furniture, other structures

Scent: (1) spraying of urine on surfaces and objects
(2) leaving feces uncovered
(3) applying sweat from footpads to trees, structures
(4) rubbing material from sebaceous glands located in chin, lips, upper eyelids, top of tail, anal sacs, scrotum on objects

Home on the Range

In contrast to territory that is guarded from intruders, home range is the area where an animal normally lives. The home range includes the cat's favorite places within his home—bed, lookout post, kitty condo—and the places outdoors that the cat frequents in his normal activities. Most outside cats have a range that consists of their yard and sometimes adjacent yards, parks, and wilderness terrain.

Cats have established pathways within their home range that connect various areas they frequent; although these trails are not marked by gravel or concrete, most cats remain constant in their use of them. Cats use smell and sight to develop mental maps of their regular route, thereby putting their beat on remote control, which saves valuable thinking time when the cat is out hunting or courting.

Neighboring cats usually have overlapping paths, and individuals may elect to avoid or to fight those who cross their path. Some cats are so regimented to using their established course that they will risk dogs, cars, and other dangers to use them.

Pita, now my mother-in-law's cat, injured the same paw, time after time, when she went under, over, or through some sort of restraining object, probably the chain-link fence, in her previous owner's yard. I saw Pita almost as often as if she lived with me, contributing, I'm sure, to her owners giving her to my mother-in-law. Since living at Grandma's, Pita hasn't reinjured the paw once; her new pathways don't involve a chain-link fence.

Cat House

Indoor cats also have pathways and regularly established areas for sunning, lounging, playing, and eating. Pita's favorite place is the windowsill, which offers certain advantages: It is high, so the cat can look out and see what's going on; it is wide enough for her to settle into comfortably; and she is close to Grandma, who might at any time offer solace in the form of food and petting. Like Pita, most cats form an attachment to at least one location that offers height, leading manufacturers of cat cages to add perches to their design and entrepreneurs to fashion elaborate kitty trees and condos for sale to more solicitous owners.

Most cats also adopt a place that offers an enclosure or hiding place. I've known cats that claimed kitchen cabinets as their special place and became quite ingenious at opening cabinet doors, and I think every cat likes to play in an open paper sack. Sometimes, a frightened cat will become calm if you offer him the safety of the paper sack for a while.

Clients often ask me if I feel that cats kept inside all the time are deprived. According to statistics, inside cats live longer than outside cats because exposure to disease and contact with hazards such as dogs and cars are reduced or eliminated.

An owner can enrich his cat's indoor environment by installing resting perches, observation decks, hiding places, and stimulating toys. Because the well-being of cats and people is directly influenced by sunlight in the environment, provide your kitty with either a perch or deck that offers natural light. Pita's windowsill was adequate; those who live in dwellings offering limited sunlight can install full-spectrum fluorescent light bulbs that provide a similar range of illumination.

The expense and ideas for making your cat's home attractive and interesting are unlimited. I recently read in my local newspaper about a woman in California who spent $100,000 converting a garage into a six-room feline palace with French windows and other amenities. Others have designed and built fenced playgrounds and enclosures that offer cats combined indoor and outdoor living.

Cat Door

Many people cannot bear to think of their pet being confined to the house since they themselves would not like to be so isolated from the outdoors. These owners often elect to install a cat door so their pet can go back and forth at will.

Several varieties of cat doors are available, ranging from a small plastic model to an electromagnetic door that operates by small magnets located on the cat's collar. Most doors, regardless of model, come with a lock so the door can be secured when you want to make sure the cat stays inside. It is a good idea to lock the door when you are anticipating a visit from the veterinarian. Several of my reluctant patients "hit the door" when they saw, heard, or sensed by ESP or some other means that I had driven up in the Cat Clinic van.

It is usually easier to train the cat to use the door when the door remains permanently open for several days. Use care to secure the door open, as it would be counterproductive to training to have the door slam down on the kitty's head just as he elects to go through the opening for the first time. Don't force the cat through the door. Let him smell it and become familiar with it before you begin training. To begin, you might place someone on the outside, enticing the kitty to go through with a treat or sweet words. Don't hurry the cat; let him make up his own mind to go through the cat door opening. After the cat comes and goes through the opening at will, release the door and entice him through the barrier by calling his name from the other side. If you've already taught him to "come" on command, this step is a snap. Commands are covered in Chapter 7.

Tree Climbing

Climbing is an important part of the cat's locomotor behavior. A cat's claws are curved just right for grasping the rough bark of a tree trunk and going up, and cats, like kids, enjoy playing the king of the mountain game.

As in most activities, the going-up is not a problem; it's the coming-down again. Kittens, and even adult cats, may have to use trial and error before they find that turning around and backing down the tree with the claws forming grappling hooks is the best technique.

Given the opportunity, most cats eventually learn a method for coming down and will do so when it's time to sit in your lap for "Jeopardy" and munch their favorite snack. How long you wait before calling your neighbor to borrow the extension ladder depends upon your nerves and forbearance.

One of my friends admitted losing patience. "The darn cat was up there, and I was late for work," she said, "so I decided it was easier to climb the tree myself and teach him to come down rather than wait. Of course, I looked rather silly trying to climb a tree in heels and panty hose, but that's not the craziest thing I've done for my cat."

My friend's actions were somewhat on the crazy side, but what's life without a little craziness to keep us sane? I may be stretching the topic of tree climbing, but my patient Hero is a good example of a cat's climbing dexterity.

Hero was a curious kitten, and he was no stranger to trouble. After suffering a broken front leg in an altercation with the family car, Hero was trussed into a rather unwieldy splint. That'll curtail his activities until his fracture heals, thought I, Hero's doctor.

A couple of weeks after the accident I called Hero's owner to inquire about his progress. "Well, I'm not sure where he is at the moment," the owner replied. "I haven't seen him for a while, so I hope the kids haven't let him outside."

A few minutes later Hero's owner called to report that she had found the convalescent in a tree in the front yard. It was a cold Texas evening, with ice covering most surfaces, and I wonder to this day how Hero managed to climb a tree with his splint. "Where there's a will, there's a way" goes the old adage, and I believe Hero made the supreme effort to climb that tree to escape the neighborhood dog that caused his demise a year later.

High-rise Syndrome

Contrary to the old husband's tale, falling felines don't always land on their feet. Of all traumatic injuries to cats, 14 percent are the result of falls. The cat is blessed with the air-righting reflex, the ability to alter his position in midair to land feet first after a fall; however, to right himself during a fall, the cat must fall or be dropped so that his legs are in the same horizontal plane. If falling head or feet first, the cat must touch the ground to right himself.

In urban areas, numerous cats fall from windows of tall buildings each year. "High-rise syndrome" is the term used to describe the traumatic injuries suffered by cats falling two or more stories.

In an interesting study conducted at the Animal Medical Center of New York, 132 consecutive cases of cats falling from two to thirty-two floors were compared by type of injury, response to treatment, and distance and history of the fall.

Although cats have excellent night vision, more of the cats in the Animal Medical Center study fell at night than during the day. Most falls occurred in late spring through early fall, during which time windows are more likely to be open. Sixty-four percent of these free-flying felines were under three years old. Two of the 132 cats fell together, and one jumped after an insect.

The good news is that 90 percent of high-rise syndrome cats lived to meow about it, and 32 percent did not require treatment. The cat that fell thirty-two stories suffered a mild chest injury and a chipped tooth, and was released from the Animal Medical Center forty-eight hours after the accident.

Statistics of the 132 cases reveal that cats falling from five to nine floors suffered the most severe injuries. These cats are also the most likely to sustain broken bones because they tend to land on their feet and legs.

Cats, like sky divers, achieve maximum velocity during free flight. The average cat will reach a terminal speed of sixty miles per hour after falling approximately five stories. Cats falling fewer than five floors suffer fewer debilitating injuries because they fall at a slower speed. It is assumed that a cat falling from more than nine floors has time to relax and orient his legs horizontally to his body, like a flying squirrel. This position decreases falling velocity by increasing drag and allows the landing impact to be distributed over a greater portion of the cat's body.

Cats falling onto mud, water, and snow had a better chance to walk away than those falling on hard surfaces. Objects that break the fall seem to be a mixed blessing. In some cases they serve to reduce the speed of impact by breaking the cat's fall; in other instances they cause an awkward landing.

Cats suffering from high-rise syndrome have nine lives, compared to humans. The chance of a human surviving a fall of greater than six stories onto a hard surface is extremely slim. Dogs also sustain more severe injuries than cats when falling from heights.

The best protection against high-rise syndrome is prevention. Keep windows screened or closed and locked. Never assume that window bars are too narrow for your cat to squeeze through. Balconies and ledges should be off-limits to pets.

Roaming

Even if you've gone the extra mile, so to speak, to provide your cat with a safe and stimulating environment, he may still insist on leaving home without telling you where he's going or why. The answer to the where and why depends on the motivation for your cat's roaming.

Let's digress and go back to the discussion on home range. The size of that range varies from cat to cat. The female range encompasses her hunting turf, if she is allowed outdoors, while the male range is more likely to cover that area containing well-known girlfriends. Breeding males have a range three and a half times larger than females, and tomcats often stray many miles when pursuing mates.

Castration usually takes away a tom's incentive to roam; however, I have known male cats that continued to explore distance turf, even after neutering. For the most part, these cats grew up in alleys, and are addicted to the excitement and freedom of street life.

They also may reside at several different homes, returning periodically to each like a bigamist sharing his affections.

Since cats are creatures of place or territory, it is a tribute to you if your cat elects to live with you in a new home over life at his old home without you. One of my clients owned a cat named Savana that spent most of his time at the old home after his family had moved to a new house located several blocks away. My client dutifully returned to the old vacant house every day to look for Savana, and most of the time she found him there. It took several months to win Savana over to the new residence. Had a new owner occupied the house and offered the cat aid and comfort, I'm not sure who would have won his loyalties.

I recommend that owners confine the cat to the new residence for a period of up to a month. I realize that an outdoor cat may offer the same protests as a grounded teenager, but I suggest that you ignore your kitty's complaints and not give in until he is weaned from his old territory. He can be allowed outside for brief periods, sporting a current ID on his collar, when you are available to supervise.

Of course, if you've moved a thousand miles away, his sentence can be reduced to the time it takes for him to become familiar with his new home. The majority of cats cannot or will not walk the thousand miles home like Sugar, the cat cited in Chapter 3 (page 40) as an example of the mystical homing sense.

Traveling Cat

On the Road Again

Ninja, a neutered black female short-haired cat, accompanies her caretaker Marjorie and poodle Candy as they crisscross the United States in an eighteen-wheel truck. "When I first introduced Ninja to the truck, she howled for about fifty miles, but she and Candy settled right down as if they had been on the road all their lives," Marjorie told me on my first visit to vaccinate her pets at a local motel.

A modern eighteen-wheeler has most of the comforts of home, and Ninja soon had a home range mapped out in the truck. She sits up front on the passenger seat, sleeps on her favorite blanket at the foot of the bed located behind the driver's seat, or rests on the top shelf of the closet, where it is dark, quiet, and enclosed.

Ninja's litter box, food, and water are available in the truck living quarters. When the trio stops for the night or exercise, Marjorie places a leash on Ninja's collar before allowing the cat out of the truck. "Ninja loves to roll and play in the freshly mowed grass at parks," Marjorie said. "And most motels, except those in North Carolina which have no-pet clauses, are glad for my pets to stay with me in my motel room."

Marjorie is a careful owner. "I always make sure my pets are properly vaccinated," she informed me, "and I keep detailed health records and pet photos with me in the truck in case I should need them on the road."

Marjorie's pets wear identification at all times, and motel maids are alerted to the fact that pets are staying in the room, so they won't accidentally let one escape. If Marjorie leaves her pets in the truck, in the motel, or in the new home she and her husband recently purchased, she leaves a sign, similar to a "Fireman, Save My Pets" decal, outside the appropriate door. The sign lists the pets inside by description and gives names and telephone numbers to call in case of emergency. Marjorie also carries an emergency card in her wallet that says, "In case of accident, my pets are located ______. Please notify ______ at the following address and telephone number."

I cared for Marjorie's pets when they were in town, but I had little to teach her about traveling with them, for she is the expert and I the novice. As usual, I learned from my clients and patients, not the other way around.

Cat Carrier

I obtained a significant number of patients and clients for my house-call service because either the client or patient hated the ordeal of the cat carrier, but even I, at times, had to transport a cat in a cage. I also know that some owners are perfectly at ease driving with a cat curled around their neck or crouched under the car seat, but I personally think pet and owner are safer and more comfortable with the cat riding in a cat cage or carrier.

There are as many different kinds of carrying cases for cats as there are creative people to adapt, design, and build them. I've seen baskets, pillowcases, canvas bags, wooden houses, backpacks, fannypacks, metal cages, and plastic crates used for cat transport. Safety for cat and caretaker and workability are the most important considerations in choosing a carrying case for cats. Workability will depend upon need. If you are transporting a docile cat between your house and Grandma's across the street, a canvas bag or pillowcase will do just fine. If your pet is traveling from Amarillo, Texas to Paris, France aboard a jumbo jet, the carrier's design and construction are critical.

My favorite carrier for transporting cats is a metal fold-up cage with a removable waste pan. This particular cage is lightweight, even when filled with a fat cat, folds down to the size of a briefcase when stored in a vehicle, allows the cat a view of his surroundings and the person a view of the cat, is easy to clean, and never needs repair. This type of case is not suitable for airline travel or for transport during inclement weather, however, for it offers the cat too little protection. The type or types of carriers you make or purchase depend upon their intended use, and I don't think it is overly extravagant for a cat caretaker to own several varieties.

Table 10: Cat Carrier Suggestions

1. Supply a separate carrier for each cat if cats are traveling simultaneously.

2. Select a carrier large enough for the cat to stand up and turn around in but small enough so he can brace himself against carrier sides during transport.

3. Carrier walls should be strong and waterproof.

4. The carrier must provide adequate ventilation.

5. The carrier should have sturdy handles.

6. A water tray accessible from outside the carrier is desirable.

7. Choose a carrier that can be washed and sanitized easily.

8. Select a carrier with a secure latch. Do not place a lock on the carrier that requires a key or combination.

9. Cover the bottom of the carrier with absorbent material.

10. Introduce the carrier well in advance of the trip, preferably when the kitty is a young trainee of two months. Place the kitty's favorite towel or blanket in the carrier and leave the cage door open. Encourage the kitty to play, eat, and sleep in his new carrier. Eventually close the door and carry the kitty in the carrier through the house. Leave the kitty latched in the carrier for a few minutes, then work up to hours. Praise and reward good behavior. Ignore complaints.

Automobile Travel

Cats that are taken along on family trips at an early age usually adjust quickly and enjoy travel. I suggest that you introduce your kitten to the car by playing with him or offering him a snack in the car. Later, when he is used to car smells and has explored the vehicle fully, place him in his carrier before leaving the driveway in the car. Placing the cat in his carrier for a trip should become as automatic as fastening your own seat belts.

Put the carrier in a safe place, such as the backseat or luggage area of a station wagon, and go for a short excursion; around the block will suffice for the first time. Act as if the car were fun. On the next trip you might take your cat or kitten to McDonald's for a bite of cheeseburger or travel to the park and enjoy the fresh air. If you take your pet in the car only when traveling to the vet's office or grooming parlor and the cat dislikes these places, it is natural that he will come to associate the automobile with unpleasant events.

When you take the cat out of the carrier, attach a leash to his collar or harness, and make sure he is wearing a current tag that identifies him by name and gives your name, address, and telephone number.

For longer trips, potential travel problems are reduced by planning ahead. Most motels accept cats, although many require a deposit or nightly payment for each cat. Major amusement parks also offer day kennels for pets of patrons. Reservations, of course, should be made in advance.

I remember a letter from a hotel owner in the "Dear Abby" column of the local newspaper a few years back. The letter read something like this: Pets are welcome in this hotel. We never had a dog or cat that smoked in bed and set fire to the blankets. We never had a dog or cat who stole our towels, played the TV too loud, or had a noisy fight with his traveling companion. We never had a dog or cat that got drunk and broke up the furniture. So if your pet can vouch for you, you're welcome, too. The hotel manager may be on to something. I never had a cat ask me for the millionth time, Are we almost there, Mom?

If you are planning to cross state or country lines, you should also make a reservation with your pet's veterinarian well in advance of departure. (Although it is unlikely that one would be forced to produce a certificate when traveling via car, each state has different health requirements, especially regarding rabies vaccinations.) He or she will determine what will be necessary to fulfill the requirements for a health certificate, which must be issued within ten days of departure. If you are gone longer than ten days, you will need another health certificate before returning. This is also the time to discuss antimotion drugs and tranquilizers for your pet.

If your cat has been conditioned to traveling, tranquilizers are not necessary. If, however, your pet reacts like my cat C.K. with salivation and howling, a mild sedative might be helpful in relieving your cat's anxiety. Individuals, cats and people, react quite differently to tranquilizers. Therefore, I recommend that you consult your veterinarian in advance about a prescription, and make every effort to test-dose the medication for effect and side effects by giving it to your cat well in advance of your travel date.

At the time of departure, offer water but don't feed the cat within four hours of traveling. Take along a bowl and water; you might want to carry a supply of water from home to avoid upsets from a different water source. If your pet carrier is not large enough for a small litter box, stop frequently and allow the cat outside on a leash to use the litter box you've brought from home for that purpose. If your cat uses the grass, pick up fecal deposits and dispose of them in closed plastic bags. Offer the cat a light snack at midday, but wait until you have stopped for the day before feeding a meal. Never leave your pet in a closed car without air circulation or water; heat stroke can occur within minutes in a hot climate.

Take along a sufficient supply of drugs and special diets if your pet needs them. Flea and tick control products, approved and labeled for cats, should be packed if you are going camping or traveling to areas of the country where external parasites are a problem. Throw in a first-aid kit for you and your cat while you're at it.

Table 11:
Collar, Harness, and Leash Training Comments

1. Safety first. Pop-away or stretch collars allow the cat freedom if he should become caught. Never leave a cat unattended with a leash attached to his collar or harness.

2. Fit is important. The collar should be snug enough not to pull over the cat's head yet loose enough for you to place a finger between collar and neck. Cut off long collar ends. Harness measurements should be taken around the cat's chest behind his front legs.

3. Proceed step-by-step. Allow the cat to wear his collar or harness around the house until he gets used to it. This may take a few hours or a few days. Attach a short, lightweight leash to the collar and harness, and let the cat drag it around for short periods when you are available to supervise. Attach yourself to the end of the leash and follow the cat (walking the person step). Walk to the end of the leash and give the cat the command to "come." Reward the correct action. If you have not taught your cat to come, see Chapter 7 and teach this lesson first.

Air Travel

If your cat is to travel by air, make plans and reservations early. Limited numbers of small pets are allowed to travel in the cabin with passengers if they fit comfortably and safely in a carrier stored under the seat in front of you; a 16- by 21- by 8-inch carrier conforms to size regulations. American Airlines, for example, allows up to seven carriers in the cabin per flight on some of their larger aircraft. One carrier per passenger is accepted, and two cats may travel in the same carrier. Cats are not allowed out of the kennel during the flight.

Airlines require that you present your cat's health certificate at check-in; if the cat is traveling unaccompanied, two dishes for food and water must also be checked, along with feeding and watering instructions. The cat must be housed in a USDA-approved kennel. The kennel should be made of sturdy material and provide cross ventilation, with projecting rims on the outside to keep airflow from being blocked by adjacent cargo. It must be leakproof and contain absorbent material on the bottom, provide a way to offer food and water without opening the door, and have enough space for your pet to stand up and turn around. The crate should be labeled "live animal," with arrows showing the top, and provide handles for carrying. It should be marked with your name and address, your pet's name, and destination.

Pets will not be accepted for air shipment if the air temperature is less than 45 or more than 85 degrees Fahrenheit for more than forty-five minutes in either the plane cargo compartment or airport holding facilities. This regulation can be waived if your veterinarian provides a statement that your cat is acclimated to temperatures below or above these levels. Of course, the risk involved in shipping pets during extremes in weather is obvious.

Many years ago I worked at a veterinary hospital located close to the airport in New Orleans. I remember a case of an anteater, en route by air to a zoo, that was brought to the clinic for treatment of heat stroke. The poor creature was comatose upon arrival at the hospital and died soon after, but the hospital kennel man who had indulged in a night on the town prior to coming in late to work that morning had quite a shock when he found the rest of us working furiously on what he thought was a very strange-looking dog.

Although most of us dislike sending pets unaccompanied on an aircraft, it is sometimes necessary. If possible, reserve direct flights. Use airlines that hand-carry your pet's carrier onto the plane rather than place it on a conveyor belt. If you are sending a cat that is economically valuable, purchase additional liability insurance. Most airlines will pay only a stated baggage limit if your pet is lost or injured. You can declare a higher value for your pet at the time of ticket purchase and buy additional liability coverage from the airline.

Other Travel

Unless you can convince an interstate bus driver or Amtrak conductor that Fluffy is a Seeing Eye or Hearing Ear cat, you are probably going to encounter difficulty in taking a cat with you on the bus and train. The same goes for the luxury cruise ships with exotic destinations like Cancún or Anchorage.

If you are taking a large ship across the big waters, you might convince the captain to accept your cat as baggage if your destination does not preclude a cat entering its ports. Hawaii, our only state free of rabies, requires animals entering to spend 120 days in quarantine at your expense before allowing them the freedom to live on the islands. Other countries have long quarantine periods, and some refuse pets altogether.

If you possess your own boat, bus, or train, you can make your own passenger list. I have known and envied people who live on houseboats, for example. Some are caretakers to boat cats, and the cats adjust to a territory confined by the boat's limits. Cats sensible enough to adopt such adventurous caretakers must become socialized to life on a boat, just as Ninja became accustomed to life on an eighteen-wheel truck, and to a territory defined by the boat limits.

Stay-at-home Cat

Cats are creatures of place, and most, if given a choice, would rather stay at home than accompany you to a Club Med resort. If you decide to leave your feline companion at home while you travel, several options are available. You can leave kitty with family or friends, hire the neighbor's child to care for the cat in your home, procure house and/or pet-sitting services, or board your cat at a kennel.

My client Priscilla tried several of these options before finding one that proved satisfactory for her cat Tuffy. Tuffy, a long-haired male, is affectionate but sensitive and nervous. He lost several pounds and appeared frightened when Priscilla picked him up after a week at a local veterinary hospital kennel. Priscilla had a neighbor come in and care for Tuffy when she left town the next month. Tuffy developed diarrhea (cats often respond to stress with intestinal upsets), and was messy and uncomfortable by the time Priscilla returned home.

Priscilla called me before her next departure, and I referred her to Sam, an acquaintance who had recently opened a home pet-sitting service in Amarillo.

Everyone, including Tuffy, was satisfied with this arrangement. Sam liked Tuffy and Tuffy trusted Sam, and Priscilla loved knowing that Tuffy was happy and well cared for in her absence.

Cat Sitters

Pet-sitting services vary from bonded companies to individuals like Sam who operate a part-time business. Pet sitters offer once or twice a day care for your pet in his home surroundings, and may even water your houseplants and bring in the mail and newspaper.

Before leaving your cat in the hands of strangers, ask for references and check them thoroughly. Have the sitter come by to meet you and your cat well in advance of making definite reservations. Make sure that the sitter understands what you expect; a written contract is even better. You should provide emergency numbers and detailed instructions about your cat's care. These precautions should be observed if the pet sitter is a neighborhood friend who will be coming in to care for your pet.

Table 12: Information for Cat-care Providers

1. Address and telephone numbers where you can be reached en route and at your final destination.

2. Telephone numbers and addresses of friends or family members in town who have keys to your house and can make decisions about your cat's care if you cannot be contacted.

3. Telephone numbers, including emergency numbers, and address of your cat's veterinarian.

4. List of names and commands to which your cat responds.
5. Precise instructions about special diets and/or medications.
6. Location of toys, grooming tools, litter, etc.

Sleeping Over

If you are leaving your cat with friends, first ask yourself: Does my cat get along with members of my friend's family and with other pets residing at that home? Does my friend have room to comfortably accommodate my cat or cats? Is my pet well behaved enough to risk a friendship? If you can answer yes to all these questions, you are truly blessed with the right friends and pets; however, if you are thinking of leaving your solitary Persian with a family that includes three preschoolers, a parrot, two Siamese females, and a Great Dane puppy, you might reconsider.

Introduce your pet to the new environment before you leave home. Provide temporary caretakers with your cat's bed, blanket, litter box, toys, and food. If medications must be given, instructions should be explicit. I once had the neighbor of an elderly pet owner call me in tears because she had tried in vain to give Poppy, an old calico, heart pills during the time Poppy's mistress was in the hospital. The situation was easily resolved because I knew that Poppy always took her digitalis hidden in a piece of cheese.

An alternative to enlisting friends to care for your pet is provided by Pets Are Inn, a Minneapolis-based company that places your pet in a private home in your area. It provides pickup and delivery service, and will transport your pet and his toys, etc. to a residence selected from a network of screened caretakers. Pets Are Inn has grown to include franchises in twenty major U.S. cities. Information about the company can be obtained by calling 1-800-248-PETS.

Boarding Kennels

Boarding kennels catering to cats are increasing in number, and special cat wings or cats-only facilities are options for some cat owners. Regardless, careful examination of references, accommodations, and personnel is advised. Solicit recommendations from friends or a professional you trust; ask questions of the kennel staff, scrutinize kennel policy, and inspect facilities.

Kennels should require proof of vaccination against communicable diseases, provide separate quarters for each owner's cats, offer separate wards for dogs and cats, and separate ill animals from healthy ones. The kennel should have adequate ventilation and be heated and cooled for animal comfort. Cat cages should offer raised resting perches and be cleaned as often as needed. Kennel personnel should enjoy working with cats, be willing to administer special diets or medications if your pet requires them, and call or transport your cat to his veterinarian if the need should arise.

You, in turn, should not board a cat that suffers from infectious diseases or parasites. You should provide kennel personnel with pertinent information about your cat's history and care, emergency telephone numbers, and the time of your return.

Not only is communication between you and your kitty's caretakers vital, but communication between you and your cat is also important. Leave a familiar toy and a blanket or towel, preferably one with your comforting scent, for your pet. Be cheerful and positive when leaving your cat; you don't want the kitty to feel as if he is going to prison while you are going to the Bahamas.

Question

Dear Dr. Whitely,

Why isn't there insurance for pets? My three cats, all roaming toms, are constantly getting into trouble and causing me to spend time and money on them at the vet's. Don't tell me to make them inside cats or to neuter them, because I don't believe in interfering with nature's plan.

Almost Pauper in Seattle

Dear Almost,

I'm sure your veterinarian appreciates your attitude about nature's plan, for your roaming toms and their troublesome injuries are probably helping to pay for the new hospital.

You are definitely a cat owner who needs pet health insurance. Two companies currently offer policies. Medipet, a member of Fireman's Fund Insurance, and Veterinary Pet Insurance are licensed to provide pet insurance in fifty and thirty-four states, respectively.

Both companies offer plans that operate in a manner similar to human health insurance policies. Premiums depend on coverage, deductible, and age of pet, but average $75 to $100 per cat per year. Most policies cover medical treatment, surgery, diagnostic tests, and hospitalization for illness or injury of the insured pet.

Individual veterinarians or groups of veterinarians have begun to offer Health Maintenance Organization–type pet insurance plans. With HMO plans, the pet owner pays a flat fee at certain intervals and then receives future veterinary services at a reduced rate or for free. Each individual HMO decides premiums and coverage. If you elect this type of pet insurance, you must use the veterinarian or group of veterinarians offering the coverage. With Medipet and VPI, you are free to use the services of any licensed veterinarian in a state offering the plan.

For more information about insurance plans for cats, contact your veterinarian or the office of your state insurance commissioner. Information about Medipet is available by calling 1-800-345-6778 (1-800-742-5678 in Connecticut) and about VPI by calling 1-800-USA-PETS (1-800-VPI-PETS in California).

If you should change your mind about the neutering, Medipet and VPI do not cover elective surgery. In the long run, however, neutering would pay by saving money, trips to the vet's office, and the health of your cats.

My best!

H.E.W.

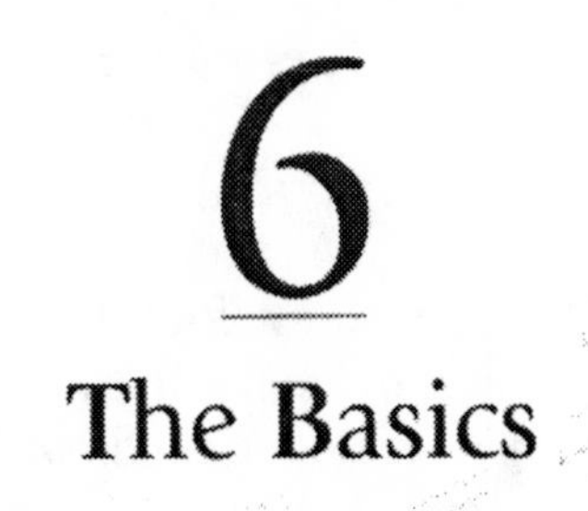

6

The Basics

Eating /Drinking

Food Preference

A newborn kitten is attracted to the nipple by scent, and smell continues to play a role in his location and selection of food as he grows into an adult. Preference for food is also influenced by the food's taste and texture, and by its familiarity.

Cats are not naturally attracted to foods with high sugar content like people and other animals, and it makes sense to refrain from giving them sugar treats. Cats are particularly sensitive to bitter tastes and cats will not eat food containing hydrogenated coconut oil because they have an enzyme in their mouth that breaks this fat and certain other fats into bitter-tasting fatty acids.

Cats that prefer dry food usually do so because they like the taste and feel of the food particles in their mouth, and they may have a preference for one particular particle shape or another. Watering a dry food usually enhances its attractiveness to dogs but not to cats. If a cat likes a dry ration, he likes it unaltered.

Kittens develop specific food preferences early in life, and their likes and dislikes are well established by the time they are six months old. Kitties develop an appetite for the type of prey their mother brought to the nest. If mother was a bird killer, kittens develop an aptitude for killing and a desire for eating birds. On the other hand, kittens raised as vegetarians may kill prey but will not eat the carcass until taught to do so by an outsider who opens the carcass, exposing them to the smell of fresh tissue.

Adult cats tend to select the type of food they ate as kittens and juveniles. One study suggests that the milk of lactating mothers contains a flavor cue and that at weaning the young exhibit a preference for the same type of diet that the mother was eating during lactation. Juveniles eating a one-food diet such as liver or tuna often develop addictions to this type of food.

People have the same tendency to choose foods they grew up with. There are usually pleasant memories associated with our favorite childhood dishes. If you want to nurture someone who is ill, offer to cook him a piece of French toast, grilled cheese sandwich, or that old standby, chicken soup, prepared just the way his mom did it.

Whether you are raising kids or kitties, it helps to expose them to a variety of foods at an early age. Kittens eating a limited menu often grow up to be picky. Morris has become rich and famous by being finicky, but for most cats being hard to please is a handicap. It is a health problem when a cat with a chronic illness refuses to eat a needed prescription diet or when a cat accustomed to a diet consisting only of liver, for example, is placed on a balanced ration by his veterinarian. Luckily, most commercial cat foods contain a number of different food items.

Although some cats initially eat more of a new diet, others turn up their nose at anything that smacks of a change in food or feeding routine. It helps to proceed slowly when enticing a reluctant kitty to accept a new food. Add 20 percent new food to the old, and when the cat accepts it add another 20 percent, and so on. The process may take a couple of months. The old theory about a cat eating when he finally gets hungry enough is true most of the time, but not all. Some cats will starve before eating what they consider an unacceptable food.

Unlike dogs, cats do not eat more in social situations. Most dogs are stimulated to eat by the sight of another dog chowing down, and many will bolt their food in an effort to be the first with the most in the food department.

Cats given a choice will select small meals, about the size of a mouse in portion and nutrients, and will eat every two to three hours, much like their hunting ancestors. An exception to this tendency to munch is seen in an occasional cat, usually female, that eats erratically, consuming little food for one to several days and then gorging the next. However, since most cats are snackers, I recommend that cat caretakers offer several small meals per day or allow free choice of dry food, if the animal is not obese.

Studies prove that cats eat and drink just as readily during the night as during the day. Cats don't need a night light to find their feeding or water bowl; they have superior night vision and whiskers that function as feelers.

One experiment showed a variation on this theme. The observed cats ate frequent small meals spaced evenly throughout the day and night, except for an absence of eating behavior during summer afternoon hours. I assume this means that the cat, being smart, is comfortably curled up somewhere, sleeping until the evening cool signals the time to awaken and get on with the business of eating and living.

Nutritional Needs

It is interesting to me that both dogs and cats belong to the taxonomic order Carnivora, but dogs have evolved into omnivores, animals that eat plants or animals, whereas cats have stayed true to their ancestral heritage by remaining flesh-eating carnivores.

The nutritional needs of cats are quite different from those of dogs, and owners should abstain from treating cats and dogs alike at feeding time. Dogs can survive quite well as vegetarians, but cats cannot.

Compared with dogs, cats have a very high protein requirement. Growing kittens require 50 percent more dietary protein than growing puppies, and adult cats require 200 percent more protein than adult dogs. The enzymes of the feline liver are geared to digesting high levels of protein and do not gear down, like those of the dog, to conserve protein when a low-protein diet is fed.

Amino acids are the building blocks of protein, and the cat requires a dietary source of the amino acid taurine. Taurine is found in ample amounts in fish, animal tissue, and feline milk, but is almost completely absent in plants. Cats receiving inadequate taurine can experience reproductive difficulties, irreversible blindness, and nervous system disorders.

Contrary to dogs, cats need a dietary source of the fatty acid arachidonic acid. This fatty acid is a constituent of animal fats but is absent in plant foods. Cats differ from dogs in their metabolism of vitamins. Dogs can convert the vitamin A precursors found in plants to vitamin A, but cats cannot. Dietary vitamin A occurs only in animal tissue. Cats also have a much higher requirement, about four times more than dogs, for the B vitamins niacin, pyridoxine, and thiamine.

Meat is the food of choice for cats living in the wild. Feral cats depending upon prey for their livelihood eat the meat of the carcass first and only occasionally consume the vegetable matter found in the stomach and digestive tract of their kill. Most of this vegetable matter has been partially digested before being consumed by the cat. Cats can digest fresh vegetable starch, but large amounts of uncooked starchy food will cause vomiting and diarrhea. For cats, the digestibility of plant material is improved by cooking.

Cats suffering from the effects of dietary deficiencies or imbalances are often those eating a one-food ration (most commercial cat foods contain more than one food item) such as liver or those eating dog food. Dogs often prefer cat food because it contains more protein and fat than dog food, and they may gobble up the cat food, leaving the cat to eat the dog food. Cats fed a vegetarian diet by a well-meaning but misinformed vegetarian caretaker may also experience similar dietary deprivations.

Drinking

When an adult cat drinks, he curls his tongue backward to form a ladle to lift liquid into his mouth. The cat takes four or five laps before swallowing. Some cats dip a paw into the water and then suck the moisture from the paw.

If fed a commercial dry cat food, cats will consume approximately two times more water than food. This is roughly the same proportion of dry matter to wet obtained by eating a mouse, which is two-thirds water. Since canned cat foods are about 75 percent water, cats eating this type of diet may consume very little water.

Cats are believed to be descended from a small desert cat native to north Africa, and as descendants of a desert animal they drink little and conserve water by producing concentrated urine. This low thirst drive may serve the cat well if he gets lost in the desert, but giving up drinking water also contributes to the health problems of cats prone to urinary tract disease. Cats, even those eating canned diets, should be encouraged to drink water.

Cats can be picky as to where and when they drink. Some demand meticulously clean water bowls and water, whereas others seem to prefer water left around for a few days. Others seem interested in water located in the toilet bowl or dripping from the kitchen faucet. Within reason, give the cat what he wants.

Because precocious kittens often fail to perceive their limits, they may tumble into an open commode when trying to drink from the bowl. It makes sense to keep the toilet lid down when training youngsters or when the bowl is filled with a toxic deodorizing cleaner.

Table 13: Feeding Do's and Don'ts

Do provide separate food and water bowls for each cat.

Do place cat's food and water bowls in a location free from noise and activity.

Do separate cat's feeding bowl from the dog's and place in a location the dog cannot reach.

Do feed food appropriate for the life stage of the cat—growth diet for kittens, juveniles, and lactating and pregnant queens; maintenance diet for normal adults; light diet for pudgy, sedentary cats.

Do provide fresh water.

Don't place cat's feeding and drinking bowls adjacent to his litter box; cats don't appreciate eating in their bathroom.

Don't supplement a balanced cat food with vitamins and minerals unless prescribed by pet's veterinarian. An excess of certain vitamins and minerals can cause health problems.

Don't feed a diet consisting of a single food item, because cats can become addicted to this food. Organ meats such as liver and kidney should not make up more than 25 percent of total dietary ingredients.

Don't feed excessive amounts of milk. Adult cats lack adequate enzymes to digest milk sugar, resulting in diarrhea.

Eating Disorders

Obesity

Obesity is the most common nutritional disease of cats, and as more cats lead pampered, sedentary lives, its incidence will no doubt increase. The same thing is happening to us: Researchers report 23 percent of Americans are overweight, up two percentage points from five years ago. Like people, cats become fat when their intake of dietary energy in the form of calories exceeds their body's energy needs; the surplus is stored as fat. In a small animal like the cat, caloric excess of only one percent can result in an animal weighing 25 percent over optimum weight several years later.

An animal is considered obese if his body weight is more than 15 percent over ideal weight. Most cats weighing more than twelve pounds are obese. Joseph, one of the heaviest cats on record, weighed forty-eight pounds. Skinfold thickness does not correlate to obesity in cats as it does in people, because the cat's skin is loosely attached and easily lifted from underlying tissue. Excess flab in cats is more commonly found over the ribs and under the ventral abdomen.

Obesity in humans and animals is associated with either an increase in the number of fat cells or an increase in the size of existing fat cells. Fat cell numbers increase most noticeably during the growth stage of life, and obesity at this time in life will predispose an individual to a lifetime battle of the bulge. Fortunately, few kittens experience the weight gain that would cause fat cell numbers to increase, and obesity caused by an increase in the size of fat cells in later life is treated more easily.

Factors that contribute to a cat becoming overweight include inherited propensity, hormonal imbalances, old age, neutering, lack of exercise, and overeating.

Some cats and people are easy keepers. It doesn't take much food to maintain their body weight. Based on lean body mass, not body weight, food intake is less for obese individuals than for lean ones. Even after weight loss, previously obese individuals require 27 percent fewer calories to maintain optimal body weight than those who have never been obese. In other words, an obese individual's metabolism becomes very efficient, and the tendency to put on weight is ever present.

As a person who suffers from being an easy keeper, I don't believe it's fair. It would be fair, however, if I were stranded with limited food on the North Pole. Then we easy keepers would fare better than our thin colleagues.

As we age, human and cat, we tend to put on weight. Most individuals lose muscle mass and add fat at the same time that physical activity lessens. As the alternative to growing old is unacceptable to most of us, it makes sense to decrease calories and increase exercise with each birthday.

Neutering doubles the chances that a cat will become overweight. A neutered animal expends less energy fighting other toms for girlfriends, roaming in search of romance, and feeding demanding kittens. Related to this reduction of sexual energy is perhaps a tendency to boredom eating. Neutered animals also put on more weight than their intact neighbors because sexual hormones depress appetite. An owner can counterbalance this tendency of a neutered animal toward obesity by monitoring the cat's calories and increasing his exercise.

A sluggish thyroid gland and other medical problems can lead to obesity, but these conditions are rare in comparison to the psychological factors that contribute to an animal's becoming overweight. One study involving dogs showed that owners who are overweight are more likely than normal-weight owners to have overweight dogs. Do we attract pets to us that have the same physical handicaps—the diabetic owner with a diabetic cat, the allergic owner with an allergic pet, the obese owner with an overweight animal—or do obese owners foist their eating habits onto their pets?

I don't know, but I've found data comparing the personalities of humans and animals that tend toward plumpness. Studies prove that overweight people and animals are wonderful plate-cleaners. Pudgy subjects are more attentive and responsive to external food stimuli than thin ones, and less attentive or aware of being satiated or full. The sight and smell of a hot fudge sundae for me or a bowl of Whiskas treats for a cat is hard to resist even if we just ate. Fat humans and cats tend to eat more rapidly than normal-weight ones, and obese individuals tend to be more emotional or excitable than controls. Obese people and animals that have lost weight tend to act just like fat ones; therefore, weight reduction is often short-lived.

Obesity is rarely a simple problem to solve, and it helps to understand the reasons behind such a behavior as overeating. Ruckus is an overweight orange-and-white tabby that had the good fortune to adopt Jack and Jacquie, true animal lovers. Jack owns an auto body shop, and one day the starved and emaciated Ruckus was found begging at the business door. Jack offered the cat a meal, thinking he'd be on his way to do whatever he normally did as a street cat. Ruckus, however, had other ideas. He knew a soft touch when he saw one, and continued to meow and beg at the shop doors until Jack gave up and took him home to Jacquie. Ruckus, now overweight, lives to eat.

Although Ruckus is well fed, he will go to his bowl and eat as if starving each time Jacquie replenishes his supply of expensive cat food or when another cat approaches his bowl. "He gorges to the point of vomiting," Jacquie said, "and I believe that he would actually go berserk if he could not get to food."

Ruckus reminds me of an acquaintance who was a prisoner during the Korean War. When this man returned from captivity, eating was his passion, and he ate without consideration for his own health or for social mores. Jacquie feels it is kinder to let Ruckus eat and suffer the ill health resulting from obesity than to limit his food supply and have him suffer the anguish of food deprivation. I agree with Jacquie. Recovery from the deep-rooted mental trauma of both individuals should come before an attempt to change the overeating behavior. Hopefully, therapy would help the man, but I'm afraid that few cat shrinks could treat Ruckus successfully.

However, most overweight cats can be treated successfully for obesity. The answer is the same as for me—limit calories and increase exercise. Exercise, which will be addressed more fully in Chapter 7, has many benefits, including burning up calories, toning muscles, increasing cardiovascular function, and, if done in moderation, decreasing food intake. In a recent experiment, exercised rats that had free access to a highly palatable diet ate 15 percent less food and gained 43 percent less weight than those not exercised. I'm sure the same principle holds true for me or for a pudgy cat.

For some animals, exercise alone is sufficient for them to lose needed weight. For others, a diet is needed. When calories are restricted during dieting, the body compensates by lowering metabolism in an effort to conserve energy—the easy-keeper syndrome. Exercise prevents this expected drop in metabolic rate from occurring, thus providing another benefit.

There are several ways to approach dieting for cats. Because of the health risks involved, I do not recommend surgery, drugs, or fasting. What I do recommend is limiting the amount of food given or placing the animal on a reducing diet.

Prescription reducing diets offer low-fat, high-fiber formulas. Lowering fat content reduces the caloric density of the ration; fiber, for the most part indigestible, increases stool bulk and prolongs eating time. Hopefully, the cat will feel full and consume fewer calories even if he is eating the same amount of the reducing diet as he was his old food.

When attempting to restrict the amount of the cat's standard diet, reduce the cat's caloric intake to approximately 60 to 70 percent of that required for maintaining the animal at his optimum weight. It is necessary to know the pet's optimum weight and the caloric density of the diet. Since most owners are not adept at making these calculations and because every diet and exercise program should be designed for each pet's physical condition, I recommend that you consult your cat's veterinarian before instituting a treatment program for obesity.

Table 14: Weight-control Recommendations

1. Physical examination and regular monitoring of cat by a veterinarian.

2. Weigh pet, estimate ideal weight, set goal weight, and set time limit for achieving goal weight. (To weigh a cat, step on a floor scale holding the cat and have a second person read the scale. Subtract your weight from the combined weight of you and your cat. Or use a baby scale.)

3. Figure the amount of regular or reducing diet to be fed per day. Divide that amount into four or five small meals.

4. Keep cat out of room when you are preparing food and family members are eating.

5. Prohibit in-between-meal snacks. Use nonfood reward for behavior training or offer small portion of day's ration.

6. Design two ten- to fifteen-minute exercise periods per day for cat. (See Chapter 7.)

7. Offer cat a stimulating environment so he will not be tempted to eat from boredom—new toys, cat perches, car trips.

8. Offer cat extra attention in form of talking, walking, petting, massage, etc., so that he will not focus on food deprivation.

9. Weigh cat every week and record his weight.

10. After cat reaches goal weight, feed amount of food needed to maintain that weight. Continue to weigh cat weekly.

Anorexia

Anorexia will be covered extensively in Chapter 9, on stress (see pages 164–66), because a decrease in appetite and/or cessation of eating are often reactions to mental or physical stress in the cat. The feline, I have found, is particularly vulnerable to this eating disorder and is most difficult to treat, even when the inciting stress is removed or alleviated.

Excessive Sucking

Kittens, particularly those that are undernourished or orphaned, may suck the bodies of littermates or themselves to satisfy a natural nursing desire. Those body parts that offer nipplelike projections are most vulnerable—ears, tails, vulvae, and scrotums.

This behavior usually stops as the cat grows into an adult and is batted away when he approaches an unwilling feline victim. Occasionally, however, the cat continues the behavior, adapting it to the suckling of the new mother figure in his life, the owner. The kneading behavior that accompanies nursing in infants is often part of the adult behavior. The cat treads with his paws and salivates as he sucks at the neck or hair of his owner.

Treatment of kittens that suckle each other may be as easy as separating them and offering a substitute object such as a rubber toy. You might also mimic the reaction of a mother cat bent on weaning her kittens—swatting the kitten on the nose and hissing loudly in his face, the cat equivalent of "no," when he attempts nursing behavior.

Diverting an adult cat's attention with a game or toy might be helpful. If diversion doesn't help, remove the cat from the room and your presence when he attempts the behavior, and reward him with praise and petting when he interacts with you appropriately.

Plant Eating

Cats often eat tender grass growing outside in spring, and this is considered normal behavior. They may vomit afterwards, but I don't think they eat grass to induce vomiting. They eat grass, particularly when it is moist and green, because it tastes or smells good.

CATNIP–Catnip is a plant in the mint family, and about 50 percent of cats get some sort of hallucinatory high from consuming, smelling, and rubbing in it. The brown or green crushed leaves are available in pet stores, or you can grow your own.

Catnip plants grow two to three feet in height and come up yearly. This perennial loves semishade and can be started in a window box planter or outdoor garden from seed (available at pet or health-food stores). Sow in the early spring in rich soil and water regularly. Cats are attracted to the tender young shoots as they begin peeking above the soil, so be prepared to cat-proof your catnip crop until the plants are well established.

As plants mature, pinch off a top stem or strip off the leaves before they turn yellow. Dry in the shade for two to three days, crumble, and store in glass jars with lids. If your cat enjoys frolicking in the catnip, rub it on his favorite toys or scratching post, or stuff a sock with it for his playing pleasure. If your cat is unimpressed by the plant, use the dried leaves to make yourself a soothing tea.

Although catnip is not considered poisonous, I believe it can be overused. The stoned state is not conducive to mental health in people or cats.

HOUSEPLANTS–Eating plants becomes a problem if the plant is toxic, has dangerous thorns, or happens to be one of your favorite houseplants. One method of discouraging the cat from eating these plants is to make them unavailable to him while making tender greens, bought at the pet store or homegrown, available as a substitute.

Another method, which falls under the category of aversive conditioning, takes advantage of the cat's propensity for smelling what he eats. The cat is forced to smell a foul-tasting substance such as pepper sauce or menthol. Then a small amount of the substance is wiped on his gums so he can also taste it. Hopefully, the cat will think "yuck" each time he smells the substance in the future. The aversive substance is then mixed with something like shortening and spread on the objects being eaten.

Wool Chewing

The cat exhibiting this eating disorder chews or suckles articles of clothing, carpeting, or furniture containing wool. Some cats will also chew other material such as silk or nylon, others will consume almost anything made of fabric, and still others choose only clothing recently worn by the owner. In one survey of cats showing this behavior, 93 percent consumed wool, 65 percent cotton, and 54 percent synthetic fabric. This study was done in England; it would be interesting to compare these statistics with fabric-chewing cats residing in a warm place like Hawaii, where wool is a rare commodity.

Cats with Siamese breeding suffer this abnormality more often than cats of other breeds, and the behavior usually first occurs during adolescence. Some breeders of Siamese cats will delay weaning until kittens are twelve weeks of age in an attempt to prevent wool sucking in kittens from mothers that have produced past litters with this behavior. Longer nursing periods for kittens are not always preventative, as the condition is more likely hereditary than due to a lack of nursing stimulation.

Some believe the condition is due to a lack of lanolin. A reader of one of my pet-care columns called to tell me about successfully treating her wool-chewing cat by administering a daily teaspoonful of a lanolin-containing hand cream. It's probably worth a try, but this owner is the only one I've known who had success with this treatment.

As with most conditions, there is no sure cure or prevention. In cases where the cat has a fetish for certain articles, the desired clothing can be locked away. If the cat has a particular favorite, such as your Laura Ashley scarf, it may be cheaper in the long run to let him have the article in the hope that he will leave everything else alone. Other cats, however, may not be so discriminating. Aversive conditioning may be tried, but few of us like to smear lard impregnated with Tabasco or Icy Hot on our favorite blouses or nylon panties.

Tactics of a stimulative and diversionary nature might be worthwhile. Allow the cat greater access to the outdoors, offer a bone with lots of gristle and sinew attached, provide free-choice dry food, offer new toys, and increase the cat's exercise periods.

If all else fails, find a home for your cat at a nice nudist colony. Owners of fabric-chewing cats will not find this suggestion humorous, and I apologize. Wool chewers can literally destroy thousands of dollars' worth of clothing and home furnishings, and the behavior is extremely frustrating to owners. Cats that cannot be stopped must often become outside pets.

Grooming Behavior

Grooming behavior using tongue, teeth, and paws serves several functions for a cat. The most obvious is personal hygiene, and lack of grooming behavior resulting in an unkempt appearance is often the first clue that the cat is suffering some form of upset or stress.

The self-groomed cat licks, bites, and paws to clean his hair and skin, to remove burrs and external parasites such as fleas and flea dirt, and to remove loose hair and tangles. Licking also offers another physical benefit—cooling the cat during hot weather by evaporation of saliva from skin and hair. As much as 30 percent of the cat's evaporative cooling loss is accomplished in this way.

By the second week of life, kittens begin self-grooming, and at about five weeks of age mutual grooming starts. The cat disperses his personal perfume or scent from sebaceous glands located at various spots over his body as he grooms his hair and skin. When he rubs against other cats, objects, or persons, he spreads that familiar smell to places in and occupants of his personal space.

Grooming is a social event. Mothers groom newborn kittens to clean them and to make that physical contact so important to the maternal-infant bond. Mutual grooming occurs when a cat extends this favor to another, much the way friends or lovers exchange back or neck rubs. This behavior is extended to humans when a cat licks and rubs against his owner or another person he considers a friend.

Most cats spend as much as half their waking time performing some type of grooming behavior. As mentioned, some cats experiencing stress will stop or decrease grooming behavior, whereas others will groom incessantly, leading to baldness. An occasional cat licks and mutilates his body in an effort to gain attention: He knows his owner is going to cry or shout when he or she sees him pulling out tail hair. Grooming can also serve as a displacement activity much in the way we bite our fingernails when frustrated, bored, confused, or frightened. Cats that are chastised for bad behavior such as biting or spraying will often run to the corner of the room to lick and pick at their hair.

A cat spends considerable time after eating cleaning his face and oral area. Since he cannot reach most of his face using his mouth, he tends to use forepaws or hindpaws as a washcloth. The favored paw is licked several times and then wiped across the neck, back of head and ears, and finally face. Most cats are ambidextrous or left-paw oriented; only 20 percent choose a right paw for the favored grooming utensil. The anus and genital areas are also carefully groomed by cats, especially during mutual grooming and at mating time.

Sleep

Cats spend at least half their life sleeping, and bored cats probably spend most of their life sleeping. Older cats sleep less than younger ones; newborns may spend 90 percent of their time in a drowsy state. Cats sleep in short spurts—cat naps—but these sleep sessions can occur one after the other.

For sleeping, cats usually choose a place that is warm and quiet. Some like elevated beds such as the top floor of a kitty condo, and others select enclosed sleeping quarters such as an open drawer or cabinet. Cats that disturb owners at night can sometimes be enticed to accept their own bed if the owner offers a heated pad or hot water bottle for comfort. The electrical brain waves (EEG) of cats are the same when they drink milk and when they are drowsy. Apparently, this physiological effect occurs before the natural sleep-inducing qualities of warm milk take over. This factor might also be of benefit to the owner trying to induce a cat to accept his own bed: Offer a small saucer of warm milk (limit the amount because milk can cause diarrhea) in close vicinity to the pet's fancy new heated bed.

After about eight hours of being awake, a cat begins to feel the effects of sleep deprivation. Therefore, don't expect your cat to join in a twelve-hour party without suffering ill effects.

Cats often dream while sleeping. Studies prove that cats can sleep with their eyes open and while walking, but most cats do not engage in sleepwalking behavior that can accompany the dream state. The cat's brain has a system that paralyzes the muscles of locomotion so they don't act out their dreams. When this neurological mechanism is blocked with drugs, the cat will sleepwalk and act out, often aggressively.

Question

Dear Dr. Whiteley,

My cat Bad Boy nearly died recently. He kept squatting in the litter box and crying. I thought the poor thing was constipated and gave him a big dose of milk of magnesia. Then he stopped eating, and I took him to the vet. The vet said Boy had Plugged Tomcat Syndrome, caused by his cat food. Anyway, Bad Boy stayed at the vet's for three days and is now eating a special food. Why do you think this happened? Did my cat get poisoned from the cat food?

Concerned in D.C.

Dear Concerned,

Plugged Tomcat Syndrome (also called Feline Urological Syndrome) is the formation of crystals or mucous plugs that accumulate in the male cat's urethra (the canal through which urine is discharged from the bladder). When the plugs block the passage of urine, the poor cat can't urinate. The condition also occurs in female cats, but because the female cat's urethra is wider and shorter, it does not become blocked.

Early symptoms include the dribbling of urine or the frequent voiding of small quantities of blood-tinged urine. When the urethra becomes plugged, the cat squats and strains in his box but is unable to pass urine. You can feel the hard, ball-shaped urinary bladder by grasping the cat gently in the lower abdomen.

Obstructed male cats must be treated by a veterinarian immediately to prevent kidney damage, uremia (waste products in the blood), and rupture of the bladder. In most cases, the obstruction is relieved by passing a catheter into the bladder. Cats are also treated with antibiotics, urinary acidifiers (crystals are less likely to form in acid urine), fluids, and diet.

Factors that contribute to the problem are the cat's tendency to drink relatively little water and to produce concentrated urine. Other factors include infrequent urination because of a cat's reluctance to use a soiled litter box or one shared with another cat, weather extremes that make a cat hesitant to urinate outside, reduced activity and obesity, and the eating of a diet high in the mineral magnesium.

Cats with a history of Plugged Tomcat Syndrome, including females, should be fed a diet containing less than .1 percent magnesium on a dry weight basis, and be provided with fresh water and clean litter at all times. Dietary control may include a canned low-magnesium diet with salt and/or urinary acidifiers added or a nutrient-dense, low-magnesium dry cat food that produces an acid urine or a prescription diet formulated to dissolve crystals in the cat's urinary bladder.

That is a roundabout way of saying, No, the cat food didn't poison Bad Boy. However, the food probably contributed to Bad Boy's having the problem. Therefore, do not change Bad Boy's food without checking with his doctor first. The cat may be on the special diet for a couple of months only or for the rest of his life, depending upon the type of prescription diet and your veterinarian's recommendations.

Best wishes!

H.E.W.

7

Teaching

Teacher and student are interchangeable positions. Sometimes, you set out to teach your cat something, and he, in turn, teaches you.

When I remarried and moved with my girls and cat C.K. from a house in the country to town, I noticed that C.K. no longer roamed and investigated the way he had in our rural community; he became fat and lazy, and rarely went outside into the noisy city environment. It was almost impossible to coax him outdoors, and I worried about the weight he was gaining.

C.K. surprised me one day by jogging the full length of the house. We heard him, not the pitter-patter of tiny feet but the thundering sound of his eighteen-pound form, as he ran the full length of the shotgun-barrel-shaped structure that was our new home. He kept up this self-designed exercise program until his death a couple of years later. C.K. taught me that there's more than one way to slim a cat.

Actually, there's more than one way to do almost anything, including teaching and training cats or people—a point emphasized by animal trainer Karen Pryor in her book, *Don't Shoot the Dog! How to Improve Yourself and Others Through Behavioral Training.*

Training Methods

Let's use the illustration of retraining a cat fearful of his cat carrier to demonstrate different training methods. Method number one is called "Shoot the Cat" (adapted from Pryor's book title). Using this method, you never put the cat in his carrier for that hated trip to the vet's office. Never taking the cat to the vet's will prove detrimental in the long run, hence you might as well "shoot the cat." Soliciting the services of a veterinarian who makes house calls would be an alternative.

Method number two is punishment. Punishment rarely changes animal behavior because it fails to coincide with the unwanted behavior, and in most cases the animal fails to understand what he is being punished for. Pulling the cat from under the bed and stuffing him into the carrier as punishment will do little to change the cat's aversion to the carrier in the future.

Method number three is negative reinforcement—an unpleasant stimulus is removed when the animal favorably changes his behavior. You might attempt to force the cat into the carrier by crawling behind him on your hands and knees, spitting and hissing at him. As soon as he enters the carrier, you stop spitting and hissing. The cat will probably think you are the one who belongs locked up in a cage.

Method number four is positive reinforcement. With this method, correct behavior is rewarded and the pet is motivated to repeat the behavior. Purchase a new carrier. Play with your cat in the vicinity of the carrier until he becomes accustomed to the sight and smell of it; place his favorite sleeping blanket in the cage so it is readily available for naps. Each time the cat enters the carrier door, reinforce the behavior with praise, petting, or a food treat.

Reinforcement

By definition, positive reinforcement is something that occurs in conjunction with a specific action which tends to increase the probability that the action will occur again. The something that positively reinforces should be an object or action that the cat wants, such as food, petting, or praise.

Negative reinforcement is something the cat will work to avoid. Squirting a cat lying on the kitchen table with a water pistol while shouting no is a form of negative reinforcement, if you stop your recriminations and squirting the instant he jumps from the table to the floor.

Wild animals are notoriously resistant to negative reinforcement methods and, in most cases, are trained using a system of rewards—positive reinforcement. Domestic animals, on the other hand, can be trained with negative reinforcement. Pulling on the leash of a dog and kicking a horse to make him go are examples of negative reinforcement. As soon as the dog follows on the leash or the horse goes faster, the pulling or kicking ceases.

Cats act very much like wild animals when it comes to negative reinforcement. Few cats can be trained to walk on a leash by pulling them along, although they can be encouraged to stroll along with you if there's a reward in it.

It is useful to think like a cat in order to know what the cat likes and dislikes. My friend Jacquie thought she was applying positive reinforcement principles when she administered an oral antibiotic to her kitten Gorgio. As Jacquie inserted the plastic dropper containing the oral medication into the side of Gorgio's mouth, she sweet-talked and praised him. "Good Gorgio, sweet Gorgio, what a good boy Gorgio is," she crooned as the cornered cat struggled in her lap. By the time Gorgio finished his antibiotic treatment, he hated the word "Gorgio" and associated its sound with the hated medicine. All Jacquie had to do was call his name and the kitten would vamoose. Instead of a positive reinforcer for taking his medication, the sweet-sounding pronunciation of the cat's name actually became a negative reinforcer. The cat now responds to a new moniker—Shy Guy.

Rewards

Positive reinforcement is the best training method for teaching cats, and it makes sense to select treats and rewards that match the cat's personality. If your cat dislikes kissing and cuddling, select a food treat or praise. If your cat loves to be scratched behind the ears or on the stomach, this would be a good way to reinforce good performance.

When using food as a positive reinforcer, keep the size of the reinforcement as small as possible—a quarter-inch cube of meat, for instance. If the cat is particularly fond of the treat, you can go even smaller. The smaller the reward, the greater the number of reinforcements per training session that can be accomplished before the cat becomes full or bored. You should also take care to select treats that are harmless to the pet. Cats suffering from a heart condition should avoid salty foods and those that are overweight should not be given high-caloric treats. It may be helpful to change from free-choice feeding to twice-a-day meals during the training period or to schedule training before feeding time to ensure that the pet is hungry enough to trade tricks for treats.

Correct timing is vital for the success of reinforcement training. If you reinforce too early, you are bribing rather than reinforcing. You are encouraging whatever is occurring at the time you offer the reinforcement. If your cat sits by the back door and meows and you are teaching him to ring the doorbell located at the bottom of the door facing, do not wave a treat in the air in an attempt to bribe him to ring the doorbell. You are reinforcing meowing rather than ringing the doorbell. Ignore the meowing, and when he accidentally hits the doorbell with his paw, immediately open the back door and say something like "Good Kitty" as you release him to the outdoors.

Pryor, the animal trainer, suggests that reinforcing too late is the novice trainer's biggest problem. The cat sits on command, but by the time you say "Good Kitty," the cat is standing again. The poor cat thinks he is being a good kitty for standing up, rather than sitting on command.

When using a food reward to teach a beginner, you should reinforce the desired behavior each time it occurs. If you are training your cat to retrieve a ring, you must offer him a tidbit of tuna, for example, and praise each time he performs correctly. After he has mastered the trick, however, the food reward can be used intermittently and eventually phased out so that the cat will perform for praise alone.

Do not use rewards indiscriminately—don't use the phrase "Good Kitty" to express affection if you intend to use it as a positive reinforcer for training the cat to perform a trick.

Table 15: Treats for Tricks

Vitamin/mineral tablet
Bite of regular ration
Small cube of cheese
Commercial treat
Quarter-inch piece of wienie, Vienna sausage
Tidbit of cooked liver
Ground meatball
Fraction of hard-boiled egg
Bite of cooked, boneless fish
Tidbit of cooked, boneless, and skinless chicken breast

Punishment

Suppose you come home from work and find your cat has left a fecal deposit inside the front door. You open the door, step in the deposit, and yell at the cat. Then you locate him and rub his nose in the unwelcome gift before throwing him outside. This verbal and acting punishment may help you let off steam, but it will do little to train the cat to refrain from misbehavior that may have occurred hours earlier.

Cats, especially those that are young or shy, often become frightened by loud words and rough behavior on the part of their owners. Other cats may become evasive, jumping off the kitchen table when they hear you approach after lounging there in your absence.

Punishment should be administered during or within a few seconds of misbehavior. The longer the period that passes between the undesired behavior and punishment, the less likely that it will change behavior effectively and the more likely other undesirable behavior will continue.

Remote punishment follows the criterion of applying retribution at the time of unwanted action. Suppose you have a problem with the cat climbing on kitchen counters. You might carefully place spring-loaded mousetraps upside down on top of the counter. When the cat jumps to the counter, the movement will cause one or more of the traps to snap and pop in the air, hopefully scaring the cat away from the counter. It is necessary to reset the traps so that follow-up attempts to jump to the counter will elicit the same type of punishment. When the cat has learned the lesson, you can put away the traps.

Teaching Commands

Most experts suggest that cats can comprehend a vocabulary of twenty-five to fifty human words. Cats don't know word definitions, but they do associate word sounds with certain objects or actions. With patience and perseverance, you can teach your cat to respond to his name, as well as to certain verbal commands such as come, sit, and stop.

Come

Suppose you've selected a name for your kitten or cat, and you want him to "come" when called by his name. For our example,

I've chosen the name Chirper because it has two syllables and sounds short and sweet.

It makes sense to expect the cat to come when called if he anticipates a pleasant experience. Don't begin your teaching by summoning your cat for medicine and baths; that can come later. Do not use your cat's name in a loud, angry voice. No cat is going to come to an owner who yells belligerently, "Chirper, come here, you just knocked over Aunt Bessie's Ming vase!"

Use your cat's name each time you feed him, even if he's standing at your feet waiting. Call your cat to come to dinner when he is in another room. Say pleasantly, "Chirper, come." Be consistent and use the same words and tone of voice each time you summon your cat. If the cat fails to respond, go get him. Put him down beside the food bowl, say "Chirper, come," and place the food in his bowl. After Chirper is accomplished at responding to the meal summons, begin to call him to come on other occasions. Reward with a food treat; eventually replace the food treat with petting, praise, or other nonfood rewards.

Sit

Your cat can be taught to "sit" on command. Place your cat on the floor and kneel beside him, preferably before feeding time. If he becomes bored and sits, give the command to sit to coincide with the action. Use the one word "sit" for the command and reward the cat with a food tidbit when he responds appropriately. If the cat remains standing, gently push down on his hindquarters while giving the command. Reward the sitting position. Work with your cat for only short periods—five minutes at the most—so neither of you becomes bored with the process. Taper off the food rewards.

Stop

I have known people who have trained their cats to "stop" on cue. This is a valuable command to use when the cat is approaching danger or mischief. To teach "stop," you must use your hand, thrust out at right angles to your arm to make a barrier against the cat advancing. Pick the right moment, as your cat is advancing toward you, to make the stiff-armed gesture right in front of the cat while you say "Stop." When the cat responds by stopping on the word cue, reward him. Later, when the cat has learned the word, you can leave out the gesture.

Teaching Tricks

Training techniques are useful when teaching an animal to perform, and teaching tricks is a great way to spend stimulating time with your cat. He receives attention from you, the reward of performing well, and health-enhancing exercise. You, in turn, develop your people-to-cat communicative skills, sometimes, and patience, always.

When deciding which tricks to teach your cat, observe his natural inclinations and disinclinations, and plan your training accordingly. If you have a cat that is frightened of the outdoors, teaching him to ring a doorbell to go outside is fruitless, for the primary reinforcer—the delight of going outside—is missing. If you have a cat that likes to stand up on his hind legs, you can reinforce that movement to teach him to sit up and beg or perhaps stand up and turn around when you give the command, "Waltz, Chirper!"

Bubba, the Siamese that rings the doorbell to solicit the services of his owner Dorothy, sits up and begs for both attention and fried chicken. If Dorothy fails to give Bubba adequate positive reinforcement in the form of praise and chicken, Bubba applies a little punishment of his own by swatting Dorothy with an indignant paw.

A kitten learns tricks more quickly and easily than an older cat, but his attention span is shorter. A two- or three-minute training session may be all that a youngster will tolerate, but an adult may work for longer periods if the incentive is delicious or interesting.

Fetching

Cats can be taught to fetch the plastic ring from the top of a gallon milk jug. Cats are copycats, so you might start the training session by demonstrating the behavior. Get down on your hands and knees so that you are closer to the cat's level. Place the ring in your open palm and allow the cat to look at and smell it. Throw the ring a short distance away and say "Fetch" before you crawl over and retrieve it, bringing it back to where your cat is standing and watching attentively. It might be good to schedule this first training session while you and your cat are home alone; if you feel uninhibited, you can get into the spirit of the game by fetching the ring between your teeth.

Stage two of this training is throwing the ring while saying to the cat, "Fetch, Chirper!" If your cat ignores you and the ring after a couple of throws, abandon the game for now. If the cat gets the idea and returns to you with the ring, reinforce the behavior with a treat and "Good Kitty." If the cat gets the ring but takes it off to a corner instead of returning it to you, ignore the cat. Reinforce only the correct response.

Jumping Through Hoop

Jumping through a hoop is a trick that will qualify your minilion for a home circus. Appropriate-sized hoops are available at stores that sell craft and needlework items, and can be decorated creatively to give a circus effect. Place the hoop on the floor and stand on the side of the hoop opposite the cat. Coax the cat through the hoop. If he is leash trained, guide him through the hoop with the leash. As he approaches the hoop, say, "Up, Chirper!" and reward him with a treat, petting, and praise when he walks through. When the cat readily walks though the hoop, raise it a few inches from the floor. Reward the cat each time he walks through the raised hoop. If he attempts to walk around or under the hoop, move the hoop into position so the cat must go through it. If the cat manages to reach the other side without going through the hoop, withhold the reward. After several training sessions, your cat should be jumping through a hoop that is elevated a couple of feet from the floor in response to the command "up."

Ringing Doorbell

Does your cat scratch at the door to go outside? Teach him to ring a bell to summon you to open the door. Mount a bell close to the place where the cat scratches at the door. Don't open the door in your cat's presence unless the bell is rung. When the cat scratches at the door, take his paw and ring the bell. Open the door immediately after the bell sounds and send the cat outside. Ignore your cat when he scratches at the door to go out. After the cat learns to ring the bell when he wants to go out, gradually move the bell to a more desirable location, such as the door facing.

Shaking Hands

Teach your cat to respond to the command to sit. After the cat is sitting, gently lift one of his paws while saying, "Shake, Chirper." Offer a reward immediately. After practicing this procedure a few times, give the command and hold out your hand. If the cat offers his paw, reward him. If the cat doesn't respond, lift his paw, shake it, and give the reward.

Playing Dead

The aim of this trick is to have your cat lie down and remain motionless when you point your finger at him and say "Bang-bang!" Give the command to sit. Point and say the word cue, and immediately push the cat over onto his side. Reward him. Repeat the procedure until the cat is responding on his own. Lengthen the time the cat must lie still before he receives the reward.

Games

Buying complicated, expensive toys and games for your feline darling is much like buying a kid an expensive Erector Set and having him play with the box it came in. Games don't have to be fancy to be fun, but they should be easy to learn and stimulating to do.

My client Nancy's three cats have favorite games devised from objects lying around the house. Fonzie enjoys the fishing pole and responds by attempting to snare the paper "fish" on the end of the line. Punky plays hide-and-seek with a paper bag, and JoJo chases a favorite Ping-Pong ball.

**Table 16:
Games Cats Play**

Ball game: catch and chase rubber balls, Ping-Pong balls, rolling toys

Soccer: Nerf ball kicking

Batting game: toy swinging from doorknob

Soap bubble chase

Light beam chase: reflect light from objects using a mirror

Birdie: paper airplane chase

Go fish: toy swung from pole

Stuffed toy: catnip enhanced

Rubber ducky: for cats that like to swim in the bathtub

Hide-and-seek: blanket or towel tent, paper bag, open boxes

Exercise Training

Of all the anecdotal evidence of the benefits of exercise, my favorite is a story about Mona, an elephant living in the El Paso Zoo. I probably like the story because I identify with a pudgy, middle-aged, and cranky female.

Anyway, Mona ate the wrong food and goofed off in her holding pen, attacking a zoo staffer who got too close and learning only one simple trick in the thirty-five years she resided at the zoo.

Mona's figure and attitude shaped up, however, after the zoo hired a Richard Simmons–type exercise guru for the inhabitants in the person of trainer Jim Bousquet. Bousquet made Mona go for walks, lift logs, stand on a pedestal, and eat a more healthful diet. Mona shed pounds and learned eleven tricks. She stopped throwing rocks and manure, and started prancing and preening to show off her new figure and tricks for visitors.

Bousquet believes every animal should have health-enhancing things to do, and that's not such a bad idea for any of us. There's little about boredom and inactivity to recommend them for cats or people; however, an exercise program that is difficult and overly strenuous for its participants has little to offer, either. Choose activities that are fun and stimulating, as well as physically conditioning, when designing your cat's exercise program.

Of course, we don't all agree on what's fun and what's boring. Even if I had known how, I would never have instructed C.K. to take up jogging, an activity I personally deplore; however, C.K. must have thought jogging through the house a fun activity, for cats rarely choose anything just because it's good for them. Perhaps one's surroundings take on a more interesting view when sliding past at a one-foot height than from the five- or six-foot level. For owners who like to jog, this might be just the exercise program to introduce to your cat. Start by walking your cat on a leash, and increase the pace and distance with each successive training session. Beware of taking your pet into areas with physical barriers and other dangers such as cars and dogs.

Like most things, exercise should be started slowly and sensibly. Check with your cat's veterinarian before instituting a program that will drastically change your pet's activity level. If you are drastically changing your own activity level in the process of changing your pet's, you'd better check with your own physician as well. Regardless of the activity chosen, short training sessions are preferable to longer ones, especially for beginners.

Climbing Stairs

For some, the advantages of climbing stairs for exercise outweigh what I consider their lackadaisical appeal. Stairs may be readily available at home, and many are covered or indoors so bad weather is not an excuse for putting off this form of exercise.

Your cat gets more exercise from climbing stairs than you just by the nature of his stature and build. It's a good idea to start out with a traditional staircase with risers backing each step; some cats are apprehensive, at least at first, when they can see between the steps to the floor below.

Place a snug-fitting collar and a leash on your pet. Pick up your cat and place him on the step closest to the floor. Give the command to "come," and when he responds reward him with praise and a treat. Proceed to steps two and three, rewarding each accomplishment. An old training maxim is to stop when training is going well. Don't proceed to the point where the pet becomes tired or bored with the activity.

Once the cat is able to go up the entire flight of stairs without fear, walk him up and down the staircase once or twice a day for a few sessions. Eventually, you should be able to remove his leash and have him accompany you up and down the stairs. If you become tired before the cat, entice him to sprint up and down the stairs to retrieve a favorite toy or to chase a ball.

Swimming

Swimming is a wonderful form of exercise, for it takes advantage of the pet's natural buoyancy in water. Water therapy is excellent for animals suffering from arthritis and other problems affecting the bones and ligaments.

If you want your cat to join you in the pool, lake, or ocean, start his initiation into water sports early. Older animals tend to be fearful if they've never had water experience.

Begin at home by introducing your kitten to the sounds, taste, and feel of water. This can be done by playing tapes of oceans and streams, and by running water in the sink with the kitten present. Pat the kitten's face and body with water from the sink. Fill the sink with a small amount of water and entice the kitten to put his feet in. Praise every step he takes toward accepting and playing in the water. If the kitten becomes frightened, begin again at a later time.

When your kitten has progressed to the point of sitting and playing in a sink, basin, or bathtub of water, it is time to introduce him to the pool, lake, or ocean. Place a secure collar and leash on the kitten, and take care that the environment is free of distractions and dangers and that the weather is conducive to a carefree outing. Let the pet sense the new terrain by feeling, hearing, seeing, and tasting the new surroundings. On the next visit, he will be willing, hopefully, to get his feet wet. Proceed slowly, praising and encouraging each step, and stop your training after an accomplishment.

Once your kitty feels confident about getting his paws wet, lead him into shallow water, taking care that the pet can stand up. Introduce a ball or other play item. When your pet is comfortable playing in the water, pick him up and take him into deeper water. Support his weight so that he won't experience an unexpected dunking. As you release your hold, hopefully the kitten will begin paddling with his feet. Once the pet is actually swimming, move a couple of feet away from him while still holding his leash. Give him the command to "come." Reward each training step.

Supervise each water excursion, and protect your pet from insects, animals, children, boats, sharp shells, and inclement weather.

Question

Dear Dr. Whitely,

I would like my cat Charlie to have the opportunity to get some sunshine in the backyard when I am outside gardening. However, he is always underfoot, and I cannot get anything accomplished. Do you recommend that he be tethered to a tree with a rope or chain? If so, how do I train him?

Gardener in West Bend

Dear Gardener,

Tethering a cat within a fenced backyard is an option for cats trained to tolerate a harness (see harness training comments in Chapter 5, page 88.) Tethering is a great way for your cat to get the benefit of the outdoors without risk of straying onto a busy highway.

Use a nylon leash, or two nylon leads fastened together, clipped to the harness as a tether. The lead can be fastened to fence posts, sturdy trees, or rings attached along the side of the house. Allow the cat a range of ten to twenty feet. An automatic leash designed for cats is available at pet stores. This product offers a self-adjusting mechanism that allows the cat to pull the leash out to ten feet.

When introducing your cat to his new outdoor territory, stay with him to offer reassurance. Play a game or introduce a new toy. Limit his first session on the tether to five or ten minutes.

As you extend your cat's time outdoors on the tether, certain precautions are necessary. Don't tether your cat where he can become entangled in lawn furniture or equipment, or fall through the slats of a porch or deck. Don't leave your cat if you can't monitor his activities from your garden. Provide fresh water and access to shade or shelter, depending upon the weather. If your cat is not trained to eliminate outdoors, provide a small litter box.

Good luck!

H.E.W.

8

Correcting Misbehavior

As more people share their intimate environment with cats, the greater are the number of feline behavioral problems that must be understood and treated. Misbehavior is often the reason for the significant number of felines brought to veterinary hospitals and animal shelters to be put to sleep.

Contributing Factors

Several factors contribute to what we call feline misbehavior. Genetics predisposes to certain conditions. The majority of wool-chewing felines are of Siamese breeding, and high-strung, nervous Burmese are often aggressive to people or other animals. Peter Neville, an animal shrink from Great Britain, reports that the majority of housesoiling cases that he treats involve Persians.

If an inherited tendency toward an emotional or mental disorder is present, stress or perceived stress on the part of the affected person or animal often triggers misbehavior for coping with the stress.

Improper socialization when the cat was young contributes to his lack of tolerance for new situations and people. Frustration adds its effects for the cat that is bored, confined, or overcrowded. Medical conditions such as epilepsy, feline urological syndrome, and lead poisoning create their own behavioral signs.

Not least in causation are the cat's natural acts that become behavioral problems when the cat's living conditions change. A cat sharpening his nails on the outdoor chinaberry tree is grooming his claws; the same cat acting in the same way by honing his pedi-daggers on the china cabinet is misbehaving. A feral female in heat attracts mates by spraying urine on the bushes; she is misbehaving if she sprays the back screen door.

Behaviors also are interrelated. The cat that is territorial may spray and scratch the furniture, as well as fight other cats entering his domain. The infantile-behaving cat might urinate on his owner's bed, as well as suck his owner's neck or clothing.

Some authorities believe that there is an innate energy associated with each type of behavior. If a particular behavior is not expressed, the unused energy is the catalyst for an alternative behavior. An example cited is that of an intact female with unused sexual energy that shows an increased tendency toward territorial aggression. I agree that unused energy predisposes a cat to mischief, but I question if a nonspayed female cat denied sex and/or motherhood is frustrated or unfulfilled, leading to behavioral problems. Regardless, the answer to this particular example is to have the cat spayed.

Because many factors contribute to feline misbehavior, treatment is rarely straightforward or easy. In most cases, it is going to take more than a brief telephone call to your veterinarian to resolve problems. Initially, schedule two appointments—one for your pet's physical examination and one for consultation about the cat's behavioral problem. If your veterinarian is not comfortable discussing behavior problems and their treatment, ask for a referral to someone who specializes in this discipline. Most large cities and universities with colleges of veterinary medicine offer veterinarians and other professionals who treat feline behavioral problems.

Housesoiling

Depositing urine or feces outside the litter box is the most common behavioral complaint that owners make about their cats. At least 10 percent of pet cats exhibit an elimination problem at some point in their life.

The first goal for diagnosing the cause of the problem is to determine if the cat is exhibiting spraying behavior or if he is eliminating inappropriately. If the cat shows a standing posture with his hind legs straight and his tail held upright and quivering, while treading with his back legs and spraying a fine mist of urine one to two feet up on vertical surfaces such as door facings, table legs, bed ruffles, and stereo speakers, the cat is spraying. If the cat squats to deposit waste onto a horizontal surface such as the top of beds or carpet, he is defecating or urinating, depending on what is deposited.

The amount of urine sprayed is usually less than the amount urinated. A cat that urinates is relieving a full bladder; a spraying cat is marking and therefore conserves his marking material so that he can identify a greater number of vertical surfaces with his odorous calling card. In some unfortunate cases, the offending cat does all three—sprays, urinates, and defecates in locations other than his litter box.

For some owners, reporting the details of their cat's housesoiling habits calls for an investigation because they seldom observe the disgusting behavior. They find the evidence—a puddle of urine behind the kitchen table, for example.

In multi-cat homes, it may be difficult to determine which cat or cats are indulging in housesoiling. If one can determine the offender by separating the cats, this is advised.

In cases of improper urination or spraying where it has been proven impossible to find the culprit, your veterinarian can give the most likely cat a fluorescein dye by mouth or injection. The dye is invisible when diluted in urine but can be detected by an ultraviolet light. The dye is administered to the cat in late afternoon. The next morning the area of the house frequented by the cat is darkened and surfaces are scanned under the ultraviolet light for the telltale dye. This procedure is carried out for two days. If the dye shows up in the cat's litter box rather than on unsuitable surfaces, you can declare this cat innocent and proceed to the next suspect. If the evidence shows urine marks on vertical surfaces, you have not only found the culprit, you have also determined the behavior is spraying rather than urinating. Obviously, this procedure is reserved for the true detectives of the cat-owning population.

After the delinquent and the precise housesoiling action have been identified, the next step is to rule out illness as a causative factor. Your veterinarian will examine the cat and order laboratory tests. If your cat has a bladder infection, he may be urinating outside the litter box because he cannot make it there in time. The cat with diarrhea may experience a similar lack of control. Cats suffering from feline urological syndrome may dribble urine when the bladder becomes full or may, on occasion, spray. Cats with feline leukemia may show any number of unusual symptoms, including wetting, which is caused by dysfunction of the muscles controlling urination. Secondary conditions such as arthritis and weakened urinary sphincter control may make it harder for an older cat to get to his litter box in time. In these cases, housesoiling is due to disease, not improper behavior.

Occasionally, however, housesoiling behavior continues after medical problems have been resolved. The cat with diarrhea that defecated on the bathroom rug may develop a liking for the soft rug material and continue the unwanted behavior. The long-haired cat that had diarrhea in or around his litter box and soiled his feet and hair may associate his distasteful experience with the litter box and avoid it in the future. A cat that associates the burning pain of urination caused by a bladder infection with his litter box may avoid the box long after the bladder condition is cured and the pain is gone.

After disease has either been ruled out or treated successfully, the next step is to answer specific questions concerning the house-soiling behavior that will give a clue as to why the misbehavior is occurring. When did the housesoiling begin and under what circumstances? Did housesoiling commence with a change in environment such as a move to a new home, boarding, or new construction going on next door?

What about the litter box? Are you placing the cat's food and water in close proximity to his toilet? Did you go on vacation leaving an irresponsible neighbor kid to clean kitty's litter box? Have you changed litter products? Are you washing the box with a strong-smelling detergent or disinfectant? Have you changed the type of boxes or the box location? Has your cat had a frightening experience near the litter box—been medicated, had mats removed, or has a box from a nearby shelf fallen and konked the cat on the head during the act of defecation or urination? Has your teenager returned from college for the summer, triggering a multitude of visitors parading past the box while playing hard-rock music nearby?

Does your cat seem to have a location preference for his unwanted behavior—urinating in the sink, on top of clothes stacked in the hamper, your new Oriental rug, your bed? Or is he indiscriminate—going anywhere and everywhere?

Have there been changes in people or pets in and around the cat's territory—a new cat hanging around the patio door or gaining entrance to the house through the cat door, a new pet, girlfriend, or baby? Does the cat exhibit the behavior when she is in heat or he hears other tomcats singing outside?

Understanding the cause of housesoiling is a step toward curing it. For example, if you determine the cat has an aversion to the new litter, the answer may be as simple as returning to the old litter product. If you notice that your cat that goes outside to do his business has accidents during inclement weather, you can provide an indoor litter box. If, however, the cat has begun spraying in reaction to the birth of your triplets and fails to respond to behavior modification or drug therapy, the answer may be as life-changing as turning the cat outside.

Spraying

Spraying is normal marking behavior for a cat. Smelling his own unique scent on the bushes or trees of his home range makes the outdoor cat feel more secure, communicates his presence to other cats in the neighborhood, and leaves a calling card for potential mates. For an indoor cat, spraying the furniture or any new object brought into the house may be reassuring and protect against rivals.

During mating season male cats mark to attract females, and females in heat mark to attract males. Therefore, intact males and females spray more than neutered cats. Neutering is a good recommendation for the spraying cat.

Castration of males after the cat has begun to spray causes a cessation of spraying behavior in 80 percent of males of all ages and a slowing of spraying behavior in another 10 percent over the next few months. There has been much debate about early neutering (before puberty) versus adult neutering in preventing spraying behavior. Studies show that it doesn't matter: The same percentage of cats neutered early as those neutered at a later age indulge in spraying behavior.

Unfortunately, approximately 10 percent of castrated males and 5 percent of spayed females spray. Intact and neutered cats of both sexes that spray to mark territory do so as a reaction to other cats within a household, to cats indoors if the marking cat is outdoors, and vice versa, and to humans in the household.

Male cats living in households with female littermates are more likely to spray than those living with male littermates, and the greater the number of cats of either sex in a household, the greater the likelihood that one or more cats will spray. In certain multi-cat households, an arrangement of shelves and boxes may allow each cat some privacy. In other cases, finding a new home for some of the cats will do much to relieve the tension of the others.

Although owners can draw the drapes or blinds and permanently shut the cat door if outdoor cats are causing a problem for the indoor kitty, most owners opt not to move back to the old house or return the new furniture or give away the baby and cannot, therefore, remedy these causes of spraying behavior.

For cats that spray in reaction to new situations or people, reducing the size of his territory to that of one secure and unchanging room may be helpful. Later, the cat's territory can be expanded when the owner is present to offer reassurance or when the cat has been calmed by drug therapy.

The spraying behavior may also be addressed directly with remote punishment. The aim is to catch the cat in the act and administer the punishment so that the cat associates the unpleasant stimulus with the location or the behavior rather than with you. This may mean nonchalantly spraying the cat with a water pistol as he sprays or booby-trapping the top of the favored laundry pile with upside-down and cocked mousetraps.

Another approach is to make sprayed areas less attractive. Clean up the urine as soon as possible so that the cat will not be tempted to freshen up his previous work. Clean with an enzymatic or biological liquid or powder detergent; rinse and wipe dry. Do not use cleaning products containing ammonia since this odor is similar to that of urine. Cover the sprayed areas with aluminum foil or with strong-smelling products like mothballs or lemon scent. Place the cat's food and water at the sprayed location because few cats like to go potty where they eat.

When attempts to curb the cat's spraying behavior by neutering, changing the environment, or using remote punishment fail, your veterinarian may prescribe drug treatment. The most common drug treatments include oral or injectable progestins (female hormones) and oral tranquilizers such as Valium.

One injection or a few weeks of oral progestin may prove effective in suppressing spraying behavior, especially if environmental factors that incited the spraying are remedied. In other cases, the unwanted behavior resumes when the drug is withdrawn. The veterinarian and client must then decide if the treatment is worth the potential side effects of long-term or high doses of these drugs—diabetes, weight gain, and inactivity. Another drawback for owners of spraying female cats is that progestin treatment is less effective in preventing spraying (about 20 percent) than in male cats (about 50 percent).

Another commonly recommended treatment is oral Valium or other tranquilizers. The results are about the same for males and females—an elimination or marked reduction of spraying behavior in 55 percent of treated cats. The side effects in this case are weight gain, drowsiness, increased friendliness, or a don't-care attitude.

As the attending veterinarian, I could certainly live with increased friendliness or a laid-back attitude on the part of my patients. The bad news is that spraying usually resumes once the drug is withdrawn.

As with most life situations, there is no sure cure that will prevent a cat from spraying or stop the misbehavior once it starts. However, understanding why the cat misbehaves in this way and attempting to alleviate these reasons are your best bet.

Inappropriate Urination and Defecation

If the cat both urinates and defecates outside his litter box, the action is more likely to be behavioral in origin than due to disease. Litter aversion is a common cause of inappropriate urination and/or defecation. A cat may develop an aversion to litter or the litter box due to inadequate cleaning, cleaning with strong-smelling products, use of a deodorized or dusty litter, too many cats using the box, dislike for the enclosed litter box, or dislike of box location. Please see Table 8 in Chapter 4 (pages 54–55) for litter box suggestions.

Table 17: Symptoms of Litter Aversion

Failure to cover feces or urine.
Straddling the edge of the box to avoid touching the litter with his feet.
Shaking his paws after doing his business.
Crawling quickly out of the box.
Crying or meowing on his way to the box.
Digging outside the box on the floor or wall.

A cat that has an aversion to using the litter box may develop a preference for inappropriate surfaces or locations. Some cats develop a liking for the cold, smooth surfaces of sinks or bathtubs, and others become accustomed to carpeting and other material. Resolution may involve all or several of the following ideas: Make the litter box more attractive, confine the cat in close proximity to the box until he uses it, make the improper site less attractive to the cat, and retrain the cat to use the box just as you would an orphan kitten (see pages 52–53). This last method involves taking the cat to the box after he sleeps, eats, drinks, plays, etc., scratching in the litter to give him the idea, and offering positive reinforcement when he performs correctly.

To make the litter box more attractive, see the suggestions on pages 54–55. It might also be a good idea to incorporate the cat's new preferences into your goal of making the litter box irresistible. In the case of the cat that likes the sink or bathtub, make him a litter box offering cold, smooth surfaces. This can be accomplished by reducing litter in his box or leaving part of the box bare; making a new litter box out of an old porcelain bedpan, adding litter gradually after the cat begins to use it; or training the cat to use the toilet (see pages 55–56). In the case of the cat that prefers carpeting or soft rugs, add a deck around the rim of the box covered with kitty's favorite material or add small pieces of rug and carpeting to the bottom of the box covered with a small amount of litter; gradually cut away material while increasing the amount of litter covering it.

Confinement to the area of the litter box can take the form of placing the cat in a large wire cage with a clean, attractive litter box located in one end or confining the cat to a small bathroom or utility room containing his box. After the cat has used his litter box for several days, his territory is expanded when you are available to supervise. Another tactic is to place numerous boxes at locations around the house; they can be moved later to less obvious locations after the cat has been retrained to use them.

Making the old surfaces less attractive may involve leaving water in the sink or bathtub, cleaning surfaces soiled with feces and/or urine, removing soiled carpeting, covering frequented areas with a plastic drop cloth or aluminum foil, at least temporarily, and changing the perceived use of the preferred area by placing food and toys in that location.

The location of your cat's misbehavior may give you a clue to the reason for the behavior. As noted in Chapter 4 (page 52), certain dominant male cats display feces as a macho means of asserting their right to certain areas. The cat that leaves a fecal deposit at doorways or walkways may be doing so for this reason. Refrain from bringing outside animals or objects into the house via this route. Leave your shoes outside for a while. Confine the cat away from entranceways for several weeks.

As I write this chapter I am reminded of the most difficult time that C.K. and I experienced in our seven-year relationship. C.K. had a favorite resting place—my side of the king-sized water bed located in the master bedroom. One day I was lying propped up against the pillows on my side of the bed reading. C.K. jumped to the foot of the bed and meowed. I ignored him, and he suddenly peed on the pillows beside me. I was horrified. I screamed at him, jumped up from my bed, scooped up the cat, and threw him out of the room, slamming the door behind him.

Never before had C.K failed to use the litter box and so blatantly showed his disapproval of me by this disgusting action. I tried to figure out his reasoning and decided that he was upset because I had taken in a series of cats to be spayed and neutered.

The hospital where I performed surgery was located behind the house where we lived. Normally, I transported the cats for surgery in my minivan and carried them into the clinic without entering the house. However, for the month prior to the urinating incident, a lady with numerous farm cats had been leaving a caged cat on my front porch every few days. I was gradually spaying and neutering all of her farm cats.

The problem, I believed, was that I was often lazy about delivering the cats directly to the clinic. The client would ring the doorbell and hand me a carrier containing the cat. I would set the carrier down in the utility room or in the enclosed back porch while I did some chore taking from fifteen minutes to an hour. C.K. was angry about the invasion of his territory, I thought.

The spraying behavior was not repeated, but C.K. went directly into a two-month-long depression. Clinical depression in cats is not described in the books, but C.K. had every symptom. He did not venture back into my bedroom for the entire period. He moved onto a kitchen chair, where he remained, except when someone wanted the chair and would physically remove him. He had the most glum expression on his face, and all efforts to bring the cat out of his depression failed. Gradually, over the period of another month, he regained his mental health and became the old C.K. that we all knew and loved.

As I researched material for this chapter, I read Peter Neville's explanation about cats that eliminate on the bed with the owner in it. In his book, *Do Cats Need Shrinks?* he writes: "The cat is emotionally upset and having found protection in its mother figure, either tries to endorse the bond or simply overrelaxes and involuntarily urinates in relief."

C.K. was not showing the territorial marking of a dominant cat; he was expressing his anxiety to his mommy, and I overreacted, contributing to his depressed mental state. I am thankful that we resumed our old relationship before his death, and am working toward forgiving myself for my lack of understanding and poor judgment involving all my children, including C.K.

BED WETTING–Complaints about cats urinating or defecating on the bed are common. In most cases, it is the insecure cat that seeks the smell of areas frequented by the owner or of items used by the owner. When the owner is absent, the cat may urinate or defecate on these areas or objects in an effort to comfort himself by mixing his smell with that of the absent owner. The cat may develop a liking for the soft bedding and continue the behavior when the owner is present. Treatment is aimed at making bedrooms off-limits while retraining the cat to use his litter box. Confine the cat to a small room or board him while the owner is away from home for more than a couple of days.

Aggressive Misbehavior

Aggression is a word that brings to mind the hunter or soldier marching off to acquire food, spoils, and territory, or perhaps it describes those who stay at home, protecting family and country from those who would invade. Aggression aimed at protecting home and family and providing food is innate behavior, and most species could not survive without it.

Aggression in cats can be normal behavior—the aggressive

hunter, the mother protecting her kittens, the tom chasing away rivals—or abnormal—the despot, the agitated cat that redirects its anger toward innocent bystanders, the kitten attacking the ankles of children, or the cat biting the hand that pets it.

Both normal and abnormal aggressive feline behavior can prove to be problematic to owners, and aggression accounts for the second most common behavioral problem in cats. Twenty-nine percent of calls about feline misbehavior made by unhappy owners to an animal behavior clinic concerned aggressive behavior; approximately 65 percent concerned housesoiling.

Aggression Toward People

Feline aggressive behaviors directed toward people occur less frequently than those directed at other cats. However, for cat owners, biting and scratching of family members or visitors is more worrisome and represents the majority of feline aggressive complaints for which owners seek professional help.

FEAR-INDUCED AGGRESSION–Aggression against veterinarians is behavior I'm quite familiar with. Often, the aggressive behavior begins as fear and the body language described in Chapter 3 is evident —body lowered against the table or ground, ears flat against the head, tail lashing back and forth, teeth exposed, hissing or growling, pupils dilating. The posture can become even more threatening —hair standing on end, tail perpendicular to body, arched Halloween stance, or lying on side with all four feet and teeth poised to defend himself. Dog owners often mistake this last cat posture as indicative of submission as it is in the canine and are surprised to find themselves fighting off an aggressive cat.

A cat showing any of these body postures is clearly issuing a warning, and if he cannot escape he will attack. An encounter with a fearful cat is the most frequent reason that people, including veterinarians, are bitten and scratched.

A cat may show a fearful reaction to people who smell, look, or sound different from those to whom they've been socialized. Misty, the calico cat who acted with fearful aggression toward William, her owner Roberta's serious boyfriend, is an example of a cat who had not been socialized to men. Cats also react fearfully toward people who have hurt or frightened them (the veterinarian or groomer). Children who pull a cat's tail or ears may condition the cat to fear all youngsters. The animal may be startled while sleeping and lash out in fright; the animal may continue to fear the person or object that elicited the scare.

Treatment of fear-induced aggression includes exposing the cat to the fear-eliciting stimulus without inducing anxiety. Once the cat can tolerate the presence of the fear-inducing person or object without a hair-raising episode, counterconditioning principles are applied. With counterconditioning, the animal experiences emotions other than fear when associating with the feared person or object.

Since I saw my fearful patients on an infrequent basis, counterconditioning the pet to tolerate my presence and actions was too time-consuming to be practical. Instead, the owner and I had to decide if the gain of the medical procedure warranted the pain of an encounter with the cat. If I had to drag the cat out from under the bed or sofa or if a fight ensued, as it often did, the cat was even more fearful of me on the next visit.

When I opened Cat Clinic, my husband, George, bless his brave heart, often accompanied me on calls to act as cat restrainer. Sometimes, the cat would associate the hated vaccinations or other procedures with either George or my medical bag. In those cases, I remained the white-coated heroine while George or the bag became the baddy. If I showed up the next time without George or the bag, the cat acted as if nothing was amiss. If George or the bag appeared, the cat declared war.

COUNTERCONDITIONING THE AGGRESSIVE CAT—I promised in Chapter 2 that the story of Roberta, Misty, and William had a happy ending, so here goes. I recommended that William stay away from Roberta's apartment until he could be present without provoking the cat's ire. Roberta displayed clothing recently worn by William in the apartment. She played recordings of William talking so that his voice became background music during the time she and Misty spent together after work. Roberta sprinkled some of William's aftershave on her own wrists. If Misty acted aggressively by growling at Roberta, the clothing, or the tape recorder, her owner ignored the protests. After a week Misty appeared oblivious to inanimate clues of William's presence. Roberta could even talk to William over the telephone without the cat trying to knock the receiver out of her hand.

Roberta and I then decided to reinforce Misty's laissez-faire attitude by giving her a low dose of Valium. Roberta administered the tranquilizer the day before William was scheduled to visit and placed Misty in her cat carrier. The cat protested at first and then settled down for a nap. The next afternoon Roberta repeated the scenario, but this time William had been invited to the apartment.

When William rang the doorbell Misty raised up and peered out of the carrier. When the cat heard and saw William through the openings of the carrier she began to growl, but Roberta reported that it was a rather weak version of her usual ferociousness. William and Roberta ignored the cat, and Misty soon ceased her protest. The procedure was repeated for a week.

The following week Roberta administered the tranquilizer before William's visit but refrained from placing Misty in her carrier. When William arrived Misty offered a weak protest and scooted under the couch. Roberta had been instructed to withhold the cat's food for the day, and with Willliam sitting quietly in a dining-room chair, Roberta offered Misty her favorite canned food in a dish located a foot away from where William sat. Misty refused to come out from under the couch, and when William prepared to leave the apartment Roberta removed the food dish. Misty went to bed without supper that night.

Each step toward William's winning Misty's confidence built on an earlier acceptance. Soon, only William was offering Misty food, and he was finally able to pet her and even hold her for brief periods in his lap. The tranquilizers were tapered off after a month. Conditioning Misty to accept William was a slow process, but because Roberta was determined to make William a permanent part of her and Misty's life, it was finally accomplished over a two-month period.

REDIRECTED AGGRESSION—This form of feline aggression might be called breaking up the cat fight syndrome, for this is how most owners experience it. It is the most frequently diagnosed feline aggressive behavior directed toward people.

The cat is involved in aggressive behavior toward another animal, a person, or an object (lawn mower or car, for example) and someone else happens by or tries to interfere, and wham—that innocent bystander gets the brunt of the teeth and nail action.

Redirected aggression can occur without an actual confrontation. If your cat is stimulated by a dog or cat outside its window and you happen by and reach over to pet him, the cat might strike out at you. If the cat has been chased up a tree by a dog, now gone from the scene, and you try to help the cat down, he may act aggressively toward you.

Cats can remain aroused for a considerable period. One study of fourteen cats diagnosed with redirected aggression shows that six remained aroused and potentially dangerous to family members thirty minutes or longer after the attack. Enough time may have elapsed between the cat's initial upset and his redirected assault that an owner is unable to associate the attack upon himself with an earlier event and believes the aggression to be unprovoked.

Treatment of redirected aggression is directed at the inciting person, animal, or object. This may mean counterconditioning the cat to accept a person (the William and Misty story), reintroducing the new cat or finding him another home (to be covered in the section on aggression toward other cats, page 145), closing the drapes, banishing the family dog to the yard, or mowing the lawn when the cat is occupied elsewhere.

The most important thing when dealing with a cat that is aggressively aroused is to protect yourself. Do not jump in to break up the cat fight with your bare hands. Throw a blanket or net, use a stick, call animal control, try to distract the cat with a ball or toy. When the cat has calmed down and is occupied with another behavior—rubbing against your leg, grooming, eating, or napping—you can safely pick him up.

PLAY AGGRESSION–Play aggression directed toward people is most common in households with one cat, cats younger than two years of age, and households where the cat is left alone much of the day. The aggressive behavior may be extended toward all family members or only certain individuals.

Normal play behavior often mimics hunting behavior—stalking, leaping sideways toward an opponent, pouncing and biting at moving objects, which may turn out to be the owner's hands and feet. If the owner flails his hands or feet in response to the attack, the cat is stimulated and continues what seems to him to be a game. A better tactic would be to ignore him or gain his attention with a loud noise, such as a sharp blast from a small foghorn and/or a rolling toy.

Never indulge in rough swatting behavior (play-fighting) with your kitten or cat. Remember that rough play and handling during the socialization period will result in the kitten becoming wild and aggressive or timid and nervous with people.

Interact with your cat using objects such as a ball or rope, and do not play with him by using your hands or feet. If your cat has a routine to his attack—pouncing on you when you open the front door after work—keep a diversion handy. This may be a ball or other toy that you throw out as soon as you open the door, or may be a change in your routine such as using another entrance or having another family member distract the cat at the time you are to arrive. Increase your cat's exercise and play periods; redirect the cat's energies by adopting another cat to act as playmate.

PETTING AND BITING SYNDROME–Some cats will suddenly bite or scratch after sitting quietly for handling and petting in their owner's lap. The cat seems to reach a threshold of stimulation, feels trapped and vulnerable, and reacts by lashing out at the owner. This behavior is reported more commonly in tomcats, but the reason is unknown.

These cats should be handled for only brief periods and in such a manner that the cat can escape easily. Warning signs of impending misbehavior include twitching of the tail, restlessness, and fake bites. If the cat shows any of these signs, put him down immediately. Do not stroke him on sensitive areas such as the abdomen or hind legs. Male cats displaying this type of behavior may respond to castration or hormone treatment.

THE BULLY–Sometimes, a cat becomes a bully. Tom, the gray tabby that screamed at me like a panther, was such a cat. He was raised from kittenhood by a young married couple. The husband played rough-and-tumble games with the kitten, and Tom's mild nips and scratches became increasingly more serious as he grew older. The adult Tom, neutered and declawed by this time, continued to indulge in unprovoked attacks upon visitors and the wife. Finally, Tom bit the husband seriously enough for him to require fifteen stitches and a tetanus shot. Not long after the attack upon the husband, the couple sent Tom to the animal shelter and adopted a new kitten.

Was Tom's aggressive behavior innate or created by his environment? It would be hard to know without a profile of his feline parents' behavioral traits. Domestic cats are not especially prone to aggressive tendencies, as are indoor cats of Burmese ancestry. Did the cat have a brain tumor or epileptic condition that triggered these attacks? Did he suffer from hyperactivity brought on by allergies to certain foods, preservatives, or dyes in his cat food? I had no clue that these were factors.

I do know that the cat was socialized to aggressive play. He also received little discipline from the wife when the husband was away at work. Perhaps, if his upbringing had included proper socialization and discipline, he would have grown up to be less of a bully.

Rough play should never be tolerated, and a kitten that shows tendencies toward becoming a bully should be disciplined early, often, and consistently. If the kitten approaches with biting and scratching in mind, say "no" firmly and give him a firm tap using one or two fingers across the bridge of his nose. Give him plenty of exercise and exposure to outdoor activities, and reward correct behavior.

Some cats, like some people, grow up to be incorrigible in spite of what appears to be a proper upbringing and background. If a physical reason for such behavior cannot be found and treated and behavior modification fails, the only recourse for those in danger is imprisonment or capital punishment of the aggressive animal.

Aggression Toward Other Cats

TERRITORIAL AGGRESSION—Social rank and territory are discussed in Chapter 5, beginning on page 74. To reiterate, territory is the area a cat guards as his exclusive possession and which he will defend against other felines. Outside cats space themselves in such a manner as to offer solitude and a secure hunting area; indoor cats often seek a place of their own for at least part of the day.

Territorial aggression, evidenced by the aggressor hissing, growling, and chasing the cat to be banished, occurs when a new cat is introduced into the house or when resident cats of either sex reach behavioral maturity, usually at two to three years of age. The aggressive cat is not necessarily the oldest or the cat that has resided in the house the longest. Several cats that interacted well may become territorial when one more cat is added. Often the territorial cat waits outside the victim's hiding place or path and leaps out to confront him.

Proceed slowly when introducing a new kitten or cat to the resident cat or cats (see pages 30–31). If the physical stress of moving to a new residence or the absence of one cat owing to an extended illness causes territorial disputes to erupt between cats that were former friends, make introductions again as if the two were strangers. If efforts to condition cats to each other fail, you may have to provide separate living areas, allowing the cats together for brief periods when you are available to offer nonstressful activities such as play and feeding. If the cats continue to fight, the separation may have to be permanent.

AGGRESSION BETWEEN MALES—Some of the most serious ritualized fighting occurs between intact male cats fighting for access to females. Prior to fighting, the two rivals will stalk each other. The body language tells it all: The base of the tail is raised, the head is moving from side to side, hair is bristled, the ears are pulled down sideways with the back edges folded, the pupils are dilated, and the cats are growling and hissing. If one of the cats looks directly at the other, his pupils beginning to constrict, the other usually stops, as he knows an attack is imminent. The attacking cat springs forward and bites at the nape, head, and shoulders of his opponent. The defending cat throws himself on his back, holds the attacker with his front feet, and scratches with his back feet. The two cats scream and roll together. Finally, one cat breaks away and the two stand facing each other, waiting for the attacking cat to resume the fight or walk away.

By the time a fighting tom is a few years old, he looks his age and more; his huge old head and shoulders are thick with scar tissue, and his ears are torn and twisted. He has kept his local veterinarian in business suturing lacerations and opening abscesses.

Castration is effective immediately or over a period of weeks or months in 90 percent of cases of intermale fighting. Treatment of the remaining 10 percent consists of administering long-acting progestins and/or confining the fighter to the house.

SIBLING RIVALRY–In multi-cat households, cats will often play roughly together. They may chase, bat, and nip at each other, but you know that they are playing rather than fighting because the type of stalking, growling, and intense biting and scratching described for intermale aggression is missing; bites are not serious and claws are sheathed.

Occasionally, the playing will become too intense, and one of the cats will terminate the play. He does this by exhibiting defensive posture (laid-back ears and erect hair). If the other cat is not serious about his aggression, he will walk away and groom himself. If, however, the larger and more vigorous cat continues the bout until the weaker cat is seriously defending himself, you may need to step in and break up the play-fighting, but not with your hands. You distract the aggressor with a ball or rolling toy or with food.

REDIRECTED AGGRESSION–In the same way that a cat redirects his anger toward a person, he does so with other cats. He may become frightened or feel threatened by seeing a bird or new cat outside his window and attack his brother as he walks by. The arousing stimuli can be an animal, person, or disagreeable noise or odor. Treatment is again aimed at decreasing exposure to the initiating source or counterconditioning the cat to tolerate the source. People engage in this same type of behavior: The boss reprimands the employee who fights with his spouse who yells at the children who kick the cat. I assume the treatment is the same with people: Learn to handle anger or quit the job.

FEAR AGGRESSION–Orphan kittens raised without benefit of interactions with mother and siblings are more likely to grow up to be fearful of other cats than those exposed early to their own kind. One cat within a household can suddenly become fearful of another, even if they've previously gotten along well together. The cat might be resting peacefully when the child's ball accidentally bops the sleeping cat on the head or body; if another cat happens to be passing by, the startled cat may suddenly become fearful of him, exhibiting lowered body posture, laid-back ears, hissing, growling, and scratching behavior toward his former feline playmate.

Treatment is aimed at separating the cats and slowly exposing the fearful cat to the other without triggering the fear reaction. This is the same method used to condition the cat Misty to tolerate the boyfriend William.

The cats can be placed in adjacent rooms with the door tied or taped slightly open (use a small wooden block between doorframe and door). One person remains with each cat and feeds, pets, and plays with him; each cat is exposed to the sounds and smell of the other without being in direct contact. The cats can also be placed in cages separated by about twenty feet. There the cats are petted and fed, hopefully without provoking a fearful reaction. Later, the cages are moved closer together. Finally, the cats are allowed together in the same room; each is supervised by a different person. Each cat is rewarded for good behavior. It is best to proceed slowly; if a fearful response is elicited in one of the cats, you have to start the training all over again.

THE DESPOT–For the most part, cats refrain from dominant-subordinate relationships; however, an occasional cat wants to be "top cat." He does this by riding-up behavior—assumption of the mating posture over both males and females—and by insistence on eating and being petted first. If another cat challenges his right to the dominant position, a fight ensues. If the bully wins, he remains in a dominant position. However, a good dose of defeat may cure his need to be a despot.

The despot that I remember was a large, long-haired spayed female cat that had been an only-cat until her owner, a nice little old lady, adopted a small Chihuahua dog that had previously belonged to a friend who had died.

Normally, I would put a Chihuahua up against a pit bull and bet on the Chihuahua, but this poor little dog was totally dominated and tormented by this cat. The dog, shaking with fright, stayed under the living-room couch, which was too low to the ground for the fat cat to follow.

One day, the owner told me, the cat was passing in front of the sofa where the dog was hiding when the lady's walker fell over and hit the cat on top of the head. The cat cried out in pain, and the sound must have triggered some primordial instinct in the Chihuahua because the dog ran out from under the couch, barking for all he was worth. The startled cat fled to the back bedroom, followed by the dog, nipping and barking at her feet. He chased her under the bed; she then escaped him by jumping on top of the chest, a tough act for such an out-of-condition cat.

From that day forth, the cat and the dog lived with an uneasy truce. They stayed away from each other for the most part, but if the cat approached the couch where the dog hid, he would stick his head out and bark furiously. The cat then fled to the top of the chest. A dose of our own medicine is curative for what ails us, I suppose.

Destructiveness

Scratching

Scratching accomplishes several things for a cat: conditions claws, stretches front limbs, and provides visual and odoriferous clues of the cat's presence in the area. If the misbehaving cat is scratching numerous surfaces throughout the house, he is usually marking; if he scratches one or two locations only, his motivation is more often claw conditioning.

A mother cat strongly influences the scratching behavior of her offspring, and selecting kittens from queens with acceptable behavior will reduce the chances of the kitten becoming destructive with his claws. A kitten learns to retract his claws at about a month of age and begins scratching with them at about thirty-five days. This is the age to introduce pedicures (see pages 57–58) and a scratching post (pages 58–59).

Environmental factors that influence scratching behavior include the location and texture of furniture and household objects such as drapes; makeup of sleeping area; owners' schedule; access to outdoors; availability, type, and location of scratching post; other cats on the property; and training from owners.

When you are not available to supervise, confine a kitten to a scratch-proof room—one without furniture to claw or drapes to climb—or to a large cage. Provide a scratching post near the kitten's sleeping area.

When the kitten can be supervised, extend his territory. If you catch him scratching inappropriately, administer remote punishment in such a way that the kitten associates the punishment with the scratching behavior, not with you. Monitor the kitten's movements by hiding out of sight or by using an intercom or child monitor. If you see or hear the kitty scratching where he is not supposed to, spray him with a water pistol, activate a siren, or throw a noisy object (a tin can filled with pennies) at him. If you are inventive, you might rig up a punishment device (tape recording, siren) that is activated by a remote-control or motion-activated switch.

If remote punishment is too difficult, you might want to booby-trap the scratched areas with double-sided cellophane tape or netting or disagreeable odors such as mothballs, commercial cat repellents, or Tabasco sauce (see aversive conditioning, page 107). A commercial mat (Scat Mat, made by Contech Electronics) delivers a mild electric shock when the cat touches it; it can be draped over a sofa or chair and left "on" if the kitty scratches at this location only.

At the same time that you are punishing bad behavior, you should be rewarding good behavior. When the kitty uses his scratching post, let him know how very pleased you are by offering praise and treats. Offer him plenty of supervised exercise and play periods and, if circumstances allow, extend that play and exercise to an outdoor environment.

If your cat persists in being destructive with his claws and you've made a down payment on a new china cabinet, you may have to consider declawing or one of the alternatives such as nail covers (see page 60).

I didn't keep good enough records to make a statistical study, but the majority of declawing surgeries that I booked for my patients occurred in conjunction with the purchase and/or delivery of new furniture—leather couches first, leather recliners second, and water beds third. If I had wanted to increase business, I would have stood outside furniture stores and offered a business card touting low-cost cat declawing.

Oh, I know, I should be ashamed. Just put me on the same hate-mail list with ambulance-chasing lawyers.

Separation-anxiety Destruction

Cats are less likely than dogs to suffer from isolation anxiety; however, dependent cats may act out destructively in your absence. This behavior takes the form of housesoiling, scratching, turning over plants, knocking objects to the floor, and scattering bed-clothes.

Treatment is aimed at reducing the pet's anxiety. This can be accomplished by tranquilizers or by hiring a house sitter to whom the cat has been conditioned.

Another method is to positively reinforce your absence. Your leave-taking should become a ritual the cat enjoys. Make a big display of getting out the cat's favorite blanket, toys, and some treat that you make available only when you are away. Leave soothing music or a tape of your voice playing on the self-rewinding tape recorder, and give the same verbal parting command each time you go, "Goodbye, Chirper."

Plan short training excursions; a few hours is enough, and leave the cat in a small cat-proof room with the items just mentioned. If the cat has been good in your absence, reward the behavior. If he has been bad, do not yell and rub his nose in the damage. Clean up the mess as unobtrusively as possible. When you are able to leave for several hours without the cat misbehaving in your absence, extend the area to which he is allowed access. When he has behaved appropriately while separated from you, extend the time of your absence.

Jumping on Counters and Tables

A sharp "no" when you catch him in the act will probably suffice to keep kitty off counters and the kitchen table when you are present, but will do little to teach him to cease and desist in your absence.

A friend came home to find the cat had eaten the steak she had left thawing on the kitchen counter for supper. My friend was so mad that she threw the cat out the back door and vowed never to let him back in the house. That's one way to prevent this unwanted behavior.

Another way is to booby-trap the table or counter with a motion detector that emits a loud noise or with two-inch strips of double-sided tape scattered over counter and table surfaces. Until you know that the cat is cured of this bad habit, you must make the counters/table unavailable (lock the cat out), or you must booby-trap the counters/table each time you leave the kitchen unattended. Intermittent reinforcement—the cat occasionally climbing to the counter and getting the reward of a thawing filet mignon—will make this behavior only harder to prevent or cure.

Attention Seeking

Some cat owners complain that the cat wakes them up every morning at 5 a.m., and they'd like to sleep until at least seven on the weekend. The behavior is especially frustrating when the cat fails to notice that daylight savings time has gone into effect.

It is helpful to know what the cat is waking you up for. Does he want to play or receive attention or eat? Whatever he wants, he'll continue waking you up to get it if you comply with his wishes.

Predators become active one to two hours before meals, so delay early-morning activity by postponing mealtime for several hours or leave dry food out at all times. Ignore the cat when he begs for food or attention. Do not play with the cat upon waking. It will probably be a battle of wills, but you must not give in to the kitty's early-morning antics unless you like having your cat act as an alarm clock.

The same principles apply for kittens or cats that cry or meow to be picked up. If you pick the kitten up when he cries, you are reinforcing this misbehavior. It would be better to ignore the kitten's pleas for attention, picking him up only when he is quiet.

Question

Dear Dr. Whiteley,

How do you find a veterinarian who treats feline behavioral cases?

Looking for Answers in Sweet Springs

Dear Looking,

I recommend that you contact the following organization which can supply you with the names, addresses, and telephone numbers of members who reside in your area.

American Veterinary Society of Animal Behavior
Dr. Wayne Hunthausen, President
4820 Rainbow Blvd.
Westwood, KS 66205
(913) 362-2512

Best wishes for finding those answers!

H.E.W.

9

Stress

Effects on Mind, Body, Behavior

Donald, a gentleman in his late seventies, lived alone with his Siamese cat, Sugar, until he was forced to spend several weeks in the Veterans Administration hospital with a fractured hip. His niece, who cat-sat Sugar in his absence, called me because Sugar had not eaten or played since Donald had been gone.

I decided that Sugar was confused and stressed by the sudden separation from her owner. At my suggestion, the niece played a taped message from Donald to Sugar several times a day. She also left the old sweater that Donald had worn to the hospital by Sugar's sleeping blanket. These simple things seemed to relieve some of Sugar's anxiety, and she began to eat and respond to Donald's niece.

Stress is the body's response to a demand. Stress may have a physical causation, such as the stress that a cat experiences during hunting, showing, pregnancy and milk production, trauma, disease, or surgery. Stress can also be psychological in origin, like that experienced by Sugar and other pets during changes involving people, animals, places, things, and activity level.

How a cat perceives and reacts to stress is directly related to his breeding, family background, and early conditioning. A kitten exposed to varied interactions with people, animals, and objects in that crucial growing-up period will become a more resilient adult. Those deprived of adequate socialization will be more susceptible to emotional and physical stress. Therefore, a cat's reaction to stress is as special or personal as the individual perceiving it. One cat's stress is another's stimulating play or interesting adventure. One cat's boredom is another's contentment. One cat's enemy is another's friend. . . .

Caretakers, too, have a direct influence. It has been said that pets take on the mental neuroses and physical ailments of their human companions. I have seen instances where this appeared to be true—the hypochondriac with the attention-seeking cat; the nervous owner with the apprehensive cat; the asthmatic caretaker with the wheezing cat. When the bond is close, the influence between owner and pet is great. Hence, the more stress and tension we perceive and/or experience in our lives, the greater the likelihood that our pets will show the physical and emotional symptoms of that stress.

The physical symptoms of stress occur at the body's weakest link. If a pet is prone to respiratory problems, he may develop allergies or asthma. If he has a weak digestive system, the stress will trigger bouts of diarrhea and/or vomiting. The same thing happens mentally. If the cat is quiet and withdrawn, he becomes even more so under stressful conditions. If the cat is overly dependent upon his owner, he develops behavioral problems such as wool sucking or soiling the owner's bed; if he is territorial and possessive, he may mark by spraying and scratching; if he is bored, he may be more susceptible to obsessive-compulsive disorders.

Stress-related Disorders

Examples of misbehavior discussed earlier might be included in this section, for most are stress-related. Physical ailments, as well as mental ones, are stress-related because stress has been shown to lower immunity and increase the body's output of adrenaline and steroids, leading to organ dysfunction, cancer, and disease. In fact, death is stress-related; the stressors finally wear out the physical shell completely.

Stress-related Physical Signs

COUGHING–Although cats don't cough as often as dogs, they do occasionally cough to get rid of inhaled irritants and aspirated particles. This cough is a normal physiological reflex to clear the air passages. However, coughing can be a sign of more serious illness such as cardiovascular or respiratory disease.

Asthma is a respiratory disease affecting cats and people. Feline asthma is characterized by a sudden onset of coughing, which ranges from mild to severe. The cough is due to a body response causing obstruction of the airway. Stress, exercise, air pollution, and changes in ambient temperature may all contribute to an attack.

Asthma usually manifests itself initially in young cats—one to three years of age—and is most common in Siamese and Himalayans. Like most diseases with an allergic component, attacks occur most frequently during the spring and fall.

Diagnosis of feline asthma is made from the results of radiographic and laboratory tests. Long-term treatment consists of bronchodilators and corticosteroids given orally or by injection.

Feline viral rhinotracheitis is a contagious disease causing frequent and often convulsive attacks of coughing. The cat suffering from this disease has a severe case of runny eyes and nose, and immediately invokes sympathy from human caretakers remembering their own flu symptoms of the last winter. Vaccination is preventative, and prevention is a much better choice than trying to cure the disease once your pet has it. See Table 22, page 189.

Direct injury to the trachea from bite wounds inflicted in a cat fight can be another cause of sudden coughing.

Worms, particularly in young cats, can cause coughing. The larvae (an immature stage) of the cat roundworm can penetrate the intestinal wall, enter the circulatory system, and be carried to the lungs, bronchi, and trachea. A heavily infested cat may develop a mild, usually temporary, cough as microscopic-sized larvae meander through the airways. Similar signs may be seen in cats with toxoplasmosis, lungworm, and lung fluke infection. Diagnosis is based on the presence of parasite eggs in the feces, and worming the cat with appropriate medication is the standard treatment.

Although it is rare, cats occasionally become infected with the dog heartworm. Coughing may be present when the right side of the heart and pulmonary vessels are affected. Diagnosis is based on a blood test specifically for the heartworm. At present, a heartworm preventative is not recommended for cats.

Although it is not as indicative in cats as in dogs, coughing can be a sign of heart disease. Diagnostic techniques include radiography, laboratory tests, electrocardiography (EKG), and echocardiography, which uses ultrasound to visually display heart structures.

Coughing can be a significant clinical sign of serious disease. Therefore, if tender loving care, rest, good nutrition, and a warm environment with a humidifier do not bring relief to the coughing cat within a brief period, the cat should be seen by a veterinarian for diagnosis and treatment.

DIARRHEA–Like mothers of small children, cat caretakers have to become experts in treating diarrhea to survive the raising of their small charges. To be technical, diarrhea is an increase in volume, fluidity, or frequency of defecation from what is normal. Defining "normal" is difficult. The bowel movement of some cats is normally hard and of others is soft; consistency also varies with the type of food in the diet.

Evidence of loose stool is usually noticed by the cleaner of the litter box, soiled rug, or soiled fur of a long-haired cat. In the case of a short-haired cat that uses the backyard as a bathroom, detection of a problem may go unnoticed for many days.

Diarrhea is either acute or chronic. In most instances, diarrhea of short duration requires only supportive and symptomatic therapy. Restricting the food intake for twenty-four to forty-eight hours of an otherwise healthy adult cat may be all that is needed to allow the intestinal tract to heal itself. If this period of fasting controls the diarrhea, a diet of white meat such as cooked, boned chicken can be introduced in small amounts and divided into frequent meals. Other easily digestible food choices include cooked lean hamburger, cottage cheese, and soft-boiled egg.

Kaopectate may be given three to four times per day at a dosage of one milliliter for every two pounds of body weight (one teaspoon equals five milliliters). Water should be provided if the cat is not vomiting.

After treating the symptoms of loose stool, the next goal of therapy is to determine the inciting cause of the diarrhea. The most commonly diagnosed cause of human colitis (inflammation of the large intestine with accompanying diarrhea) is emotional upset. If a person has a stressful day at work or fights with a loved one, he may suffer from an upset stomach. Cats also get stomachaches in response to a change in environment or an emotional upset. It might be that the cat is frightened by the visiting grandchildren or by the dog staying indoors in the cat's territory during inclement weather. The cat's digestive tract usually settles down to normal after the grandkids return home or the dog is kicked out into the rain. If the grandkids or the dog is going to be present frequently, the cat may benefit from counterconditioning.

Changes in or tolerance to ingredients in diet also cause

diarrhea. Diarrhea in kittens often occurs as an intolerance to cow's milk. This may be due to a deficiency of the enzyme that digests milk sugar or an actual allergy to the protein in cow's milk. If you are feeding kittens formulas made with cow's milk, try replacing the cow's milk with a commercial feline milk substitute such as KMR (Pet-Ag). Adult cats may also exhibit a similar intolerance to milk products; in this case, eliminate the cat's bedtime bowl of milk. If the diarrhea is due to a rapid change in diet, reinstate the familiar food or proceed at a slower pace in introducing the new food.

If other physical symptoms accompany diarrhea or if the diarrhea continues or recurs after fasting for one or two days or after reinstating the old diet and eliminating products containing cow's milk, call your cat's veterinarian to schedule an appointment. A professional should determine if the cat's digestive problems are due to emotional or physical stress.

A clinical workup to determine physical causes of diarrhea includes a complete examination of the cat, laboratory tests, contrast radiographs, and fecal analysis for blood and parasites.

Parasitism is more likely to cause diarrhea and weight loss in kittens than in adult cats. A kitten should be checked for worms at six weeks of age and again at nine and twelve weeks if the results are negative.

Coccida, a protozoan parasite, is a classic offender for causing diarrhea, although roundworms, hookworms, and tapeworms might also be the source of the problem.

Tapeworm segments look like grains of rice and are usually visible around the anus of the cat and in the feces. The flea is the intermediate host of the common tapeworm that infects cats, and flea control, as well as worming, is the goal of treatment for this parasite.

Kittens with profuse diarrhea occasionally suffer from rectal prolapse (protrusion of the lining of the large bowel through the external body opening) caused by constant straining. This condition may be self-correcting after the diarrhea is controlled, but surgery may be necessary.

Certain antibiotic treatments cause diarrhea because they eliminate the bacterial microflora that normally inhabit the gut. When antibiotic therapy ceases, the diarrhea is usually alleviated. However, in cases of long-term antibiotic therapy, live-culture yogurt or lactobacillus-containing medications may be added to the diet to relieve this side effect.

Diarrhea, although not a disease itself, may be a clinical sign of serious disease in the cat. The classical signs of feline panleukopenia (commonly called cat distemper)—a young, dehydrated kitten suffering from diarrhea and vomiting—are thankfully seen only rarely today because of the excellent vaccines available for its prevention. A kitten should be immunized initially against this disease when the little fellow is between six and nine weeks of age. See Table 22, page 189.

Diarrhea may also signal diseases such as feline leukemia, feline infectious peritonitis, and FIV (feline immunodeficiency virus) infection, and serious organ dysfunction such as kidney or liver failure. It can also be a presenting symptom in animals suffering from food allergy, intestinal cancer, viral and bacterial infections, and poisoning.

VOMITING–A client owned a long-haired male tabby aptly named Hair Ball. I was never concerned that the fat and otherwise healthy cat had an addiction to the Bermuda grass growing in his yard or that he vomited frequently after consuming the grass or that the vomited material often contained yucky hair and bug skeletons. What I did object to was being subjected at least twice a week to a frantic telephone description of Hair Ball's upchucking behavior.

I never convinced Hair Ball's mistress that vomiting, in this case, was a logical consequence of grooming and of eating indigestible insects, and that Hair Ball probably stimulated the act by eating grass. An old wives' tale suggests that eating grass can produce mild emesis (vomiting) in dogs and cats, and as an old wife, I believe we know what we're talking about, at least some of the time.

Vomiting is also a natural consequence of bolting food or engorging the stomach with food. Remember the example of the formerly deprived Ruckus, now obese; he eats to the point of throwing up. A dominant cat may eat too much or too rapidly in response to a new cat in the household. One remedy for cats that eat too fast or too much is to limit the food to small amounts divided into several small meals. In the case of competing cats, separate them at feeding time.

Cats vomit readily. In many cases, the cause may not be clear, but the cat recovers within a few days. An apparently healthy cat that has a bout of vomiting may be treated at home. Withdraw all food and water for twenty-four hours. Provide water in the form of ice cubes. Once the vomiting has stopped, feed a small amount of meat or chicken broth every couple of hours. Strained-meat baby food or cooked, cubed chicken breast can be added. If this is tolerated, add cottage cheese and/or soft-boiled eggs. If vomiting has ceased for forty-eight hours, small quantities of regular diet can be fed every other feeding until the patient is back to normal. Do not exercise your pet immediately after eating.

Vomiting can be divided into three recognizable stages: nausea, retching, and vomiting. Nausea, the urge to vomit, is characterized in cats by salivation, licking of the lips, and repeated swallowing. Retching, the effort to vomit, may follow rapidly. As stomach contents are brought up and out of the mouth, vomiting occurs.

The vomiting center in the brain ultimately controls the act. Fear, stress, or excitement can induce vomiting in animals and people.

Motion sickness, brought on by car travel, for example, or ear infections, both of which cause changes within the semicircular canals of the cat's ears, can also stimulate the vomiting center. In the case of motion sickness, tranquilizers to reduce anxiety and antimotion drugs may help reduce symptoms in animals that cannot be conditioned to travel in car, boat, or airplane.

Most neurological pathways to the vomiting center in the brain are located in the abdominal organs. Overstretching the stomach in a cat that eats too much of a new brand of cat food or the presence of an obstructive foreign body such as a large hair ball will trigger the vomiting center.

Regurgitation and vomiting are different actions. It is often difficult to differentiate between the two, but it can be important for proper diagnosis of a problem. Vomiting is usually characterized by marked effort; regurgitation requires little effort. Vomited material may be partially digested; regurgitated food, which is from the esophagus, not the stomach, is undigested. Regurgitated food is sausage shaped; vomited food has less form. Cats rarely attempt to eat vomited food but might eat regurgitated material.

An accurate record of each episode will be useful to help you or your cat's veterinarian differentiate between vomiting and regurgitation and to arrive at a diagnosis. Consider the following questions. How long after eating does the episode occur? How far is the vomitus propelled? Did regurgitation follow gagging or coughing in a cat with a respiratory infection? Is any material present in the vomit, such as hair balls or roundworms? Is there an odor? Is there blood? If so, is it bright red or dark?

Some cats regurgitate undigested food within a few minutes of eating a meal. Many of these cats are temperamental and nervous individuals; when physical examination of the cat and laboratory tests fail to reveal a cause, tranquilizers and the feeding of small, frequent meals may help.

Persistent vomiting or regurgitation in kittens may have a more serious cause. Pyloric dysfunction is a congenital condition that causes these signs. The pylorus is a sphincter muscle between the stomach and small intestine. A malformed pylorus does not allow the stomach to empty properly and causes vomiting. Barium X rays usually confirm this diagnosis.

A dilated esophagus (megaesophagus) may be associated with pyloric dysfunction or it can be a separate disorder. In the latter case, the cardia, a sphincter between the esophagus and the stomach, does not allow food to empty properly from the esophagus into the stomach, and kittens tend to regurgitate the undigested food remaining in the esophagus. This malformation of the digestive tract is diagnosed primarily on the basis of barium radiograph results.

Feeding frequent small meals of semiliquid or pureed food may help the kitten or cat suffering from these disorders. Holding the kitten in an upright position for feeding will allow gravity to help the food pass through the esophagus into the stomach.

Vomiting can be a sign of a serious disease. Cats suffering with lymphosarcoma, a cancer related to feline leukemia, may exhibit vomiting as an early sign. Vomiting is seen in young cats with panleukopenia and older cats with liver and kidney disease. Vomiting can occur in cases of blood poisoning that are due to abscesses or uterine infection and with intestinal obstruction.

Vomiting is a symptom, not a disease. It can be the sign of a physic upset, the ingestion of too much food, a clue to a congenital malformation, or an indication of serious disease. Other signs such as depression, dehydration, fever, and diarrhea provide a basis for making a decision about the seriousness of the condition. In a small, weak kitten, vomiting and/or diarrhea, regardless of the cause, may be the straw that broke the camel's back.

Knowing when to phone the veterinarian and when to give Mother Nature a chance is a judgment call. The vomiting cat may be a temporary nuisance or a very sick animal that requires immediate emergency veterinary care. In the case of a depressed, vomiting cat, it is always better to err on the side of caution.

ANOREXIA–An animal normally eats to fulfill his caloric needs. Anorexia, the loss of desire to eat before meeting those caloric needs, occurs when an animal experiences medical problems such as disease, trauma from injury or surgery, and cancer. Medical problems and therapies affecting a cat's sense of smell and taste are especially stressing to his appetite. A cat experiencing emotional stress—from pain or fear, for example—is also vulnerable to anorexia.

Stimulating the appetite of a stressed cat is very important, regardless of the cause, because undernutrition leads to ineffective wound healing, muscle weakness, organ dysfunction, and decreased immunological response. If an animal refuses to eat for emotional reasons, he is setting himself up for illness.

Cats are particularly hard-boiled when it comes to eating. It seems to me that an occasional cat will refuse to eat out of stubbornness. Anorexia nervosa in humans has a complex emotional causation, and control over body and eating is a factor. Although anorexia nervosa is not described in cats, I believe a form of this disease exists, and the poor cat so affected will elect starvation when forced to change food or environment such as occurs during hospitalization or boarding.

For most cats, however, anorexia is a temporary disorder, amenable to good nutrition and to tender, loving care. The most important nutrient is water. An animal can stand the loss of all his body stores of fat and carbohydrate and half his protein, but a sudden loss of 10 to 15 percent of his body water can result in death. Therefore, water should always be readily available and in the form the cat prefers. If he likes to drink his water from a dripping faucet, fix a dripping source of water next to the cat's sickbed. Providing electrolytes (salts) with the water may be helpful for a cat suffering from anorexia; your veterinarian will prescribe electrolytes if needed, including veterinary brands and others like Pedialyte (or Gatorade), which are available over the counter at pharmacies and grocery stores.

Here's an anecdote about treating animals suffering from dehydration. In the early eighties, when canine parvovirus was newly discovered and killing young dogs at an alarming rate, I attended a conference devoted to this disease. One of my colleagues from England accompanied me to a session led by a veterinarian with a strong Southern drawl. The Southern veterinarian recommended that Gatorade be administered orally to treat the severe dehydration that parvo-infected puppies experience. My English friend knew that I was Southern and turned to me for an explanation. "Where," he asked, "do you think we are going to find a source of those alligator eggs in England?"

Gatorade or 'gator eggs—what will entice a cat to eat? For some, it's that favorite boiled chicken breast; for others, it's one of those expensive gourmet canned foods, and for still others, it will be a well-balanced high-protein, high-fat prescription diet such as Prescription Diet feline p/d or c/d (Hill's Pet Products), available at veterinary hospitals. Hill's recently marketed (through veterinarians) Prescription Diet a/d, a high-protein, high-caloric diet formulated for sick and convalescing dogs and cats. The food is packaged in a syringe for ease of administration. Other liquid diets such as CliniCare (Pet-Ag) for sick animals are available at veterinary hospitals.

Although not a balanced diet, human baby food, especially the strained meat and egg yolks, may be easy for the cat to digest. High-caloric pastes, available at veterinary hospitals, are easy to administer to the cat; he will usually lick the gel off the side of his lips or paw. Like baby food, the oral caloric supplements are not balanced and are useful only for short-term treatment.

Sometimes, it is how, rather than what, food is offered. Providing the cat a quiet, warm, and safe environment free of other cats with his favorite sleeping blanket and food and water within easy reach may be helpful. Warming his food to room temperature may entice him to smell and taste a bite. Put a small bite-sized portion of food directly into his mouth. If he is suffering from an upper respiratory problem, clean his nose and humidify his air; instillation of a 5 percent solution of saline nose drops may help him breathe and smell, and therefore taste better.

If the anorexic cat is hospitalized, ask his veterinarian if you may come by once or twice a day to feed him. Soothing words and petting from someone the cat trusts and loves may convince him to take that first bite.

That first bite is important. I have had patients that continued to eat on their own if I could just convince them to take the first bite. It's as if that stubborn reluctance to eat was broken, or perhaps it was finally smelling and tasting a bite of food that triggered the appetite center in the brain.

Interestingly, a minute amount of the tranquilizer Valium injected intravenously will sometimes stimulate a noneating cat to take that first bite. When it works, the Valium treatment is dramatic: The cat walks immediately over to the food and begins eating. I'm not sure how it works, but I've observed its effectiveness when nothing else I've tried to entice or force the cat to eat did. Oral Valium is also an appetite stimulant, which is one of the reasons that cats placed on this tranquilizer for treatment of behavioral problems often become obese.

Valium is not a panacea and is useful only for short-term treatment of anorexic cats. Finding and treating the cause of the anorexia is paramount in importance. If the cat is ill, the disease or organ dysfunction must be treated. If the cat is experiencing emotional stress like Sugar when her owner was hospitalized, that stress must also be alleviated, if possible.

Stress-related Mental Ailments

OBSESSIVE-COMPULSIVE DISORDER—Obsessive-compulsive disorder is a popular diagnosis in the human health field, affecting up to 3.3 percent of the U.S. population. Someone who washes his hands over and over again throughout the day and night suffers from this disorder. Freud defines the behavior as follows: "The patient is impelled to perform actions which not only afford him no pleasure but from which he is powerless to desist."

Cats often develop similar compulsions, especially when stressed. MeeMee, a female Siamese that lived with a female sibling and a middle-aged couple, developed an obsessive-compulsive disorder when a third cat was adopted.

MeeMee's caretakers had decided to adopt the young calico cat that resided next door. The calico, primarily an outdoor cat, was accident-prone and always had a lump, bump, or laceration that needed attending; this was the reason the concerned neighbors decided that the cat needed closer attention than the previous owner, a flight attendant, was able to give her.

The calico continued to share the two yards and to get into trouble. Not long after she was living at MeeMee's house (sleeping in a cat house on the patio and eating on the enclosed back porch), the calico received a wound that required twenty sutures. I advised the new owner to keep the calico indoors and inactive until the wound healed.

In the meantime MeeMee started a very unusual sequence of behaviors: She looked around with a weird look in her eyes, swung her head to one side, and licked at her left paw in the sort of rhythm that you could set a metronome by. Nothing seemed to distract her when she entered one of her compulsive licking periods, and the licking focused on the same paw in the same place and in the same pattern—lick, lick, lick.

MeeMee's new owner called me, distraught over both MeeMee's behavior and the fact that the calico had chewed out several of her stitches and would require resuturing. I patched the ever-problematic calico back up and searched for a physical cause for MeeMee's symptoms.

MeeMee's ears and skin appeared healthy and laboratory tests were normal. I knew from previous visits that MeeMee was an extremely nervous cat, and I suggested to her owner that we tranquilize MeeMee while the calico was housebound. Voilà! Valium saved the day. As long as MeeMee was drugged, she refrained from the behavior; when the calico recovered and went back outside, we tapered MeeMee off the tranquilizer. When the calico had another mishap, and she often did, we had to put MeeMee back on the tranquilizer to prevent the compulsive disorder.

All cats indulge in displacement grooming, including licking behavior, to a certain extent. Something triggers the cat to feel internal confusion and conflict; instead of reacting to whatever is causing the conflict, the cat will stop to lick his paw or rub it over his face. C.K. did this after I scolded him about jumping on the kitchen counter: He would jump down, look confused, and groom himself.

Some cats begin excessive licking behavior to rid their skin of parasites and continue the behavior beyond normal limits. Other cats begin to lick and pull hair because of nervousness much the way a person might bite his fingernails (Siamese and Abyssinians are well represented in the category), others because of boredom, and some to gain the attention from their owners that they know the excessive behavior will bring. When the behavior becomes repetitive and extreme, we classify it as an obsessive-compulsive disorder.

I think that people and animals desire a safe and predictable environment, and lack of predictability and control can be extremely stressful to certain individuals. When the animal becomes highly aroused because of stress-inducing factors and cannot control or avoid the source of the stress, the resulting behavior can develop a repetitive rhythm, designed to soothe and take one's mind off the problem. Later, this behavior is indulged in during any period of high arousal. As a coping method, compulsive behavior fails: Research shows that obsessive-compulsive actions do not change the body's physiological reactions to stress.

Compulsive behaviors that were expressed briefly early in life are often resorted to later when the individual is stressed. I once knew a young woman who pulled her own hair out when she began taking classes I taught at a local junior college. I didn't know that the girl was mutilating herself until her sister called to explain: The girl had begun pulling out her own hair during infancy, and the behavior manifested itself in times of extreme stress.

Cats, too, compulsively self-mutilate. The extremes range from hair pulling, usually in the area between thighs and lower abdomen, to biting the tail or feet. Sometimes, the damage to the part bitten is so bad that amputation is required. Compulsive disorders can take many forms. Wool and fabric chewing, discussed in Chapter 6 (pages 107–108), can be considered a form of obsessive-compulsive behavior, as can certain forms of hallucinatory behaviors such as are seen in rolling skin disease. See Table 18 for clinical signs manifested by cats suffering from these disorders. Individual cases may show one or many of these symptoms.

Treatment of obsessive-compulsive disorders is aimed at identifying and removing the cause of conflict. Sending the calico back to her original home would be the method of choice for treating MeeMee's obsessive behavior. Another method would be to countercondition MeeMee to tolerate the calico (same approach as retraining cats showing fearful aggression, pages 151–53). If boredom or attention seeking plays a part, provide a more stimulating environment and ignore attention-seeking behavior.

Train an incompatible behavior: Give the command for the cat to "come" or train him to sit in a chair on command. It is difficult for the cat to lick while he is performing. The very best would be to teach the compulsive cat to do the family ironing: You'd be training an incompatible behavior and taking advantage of the cat's natural inclination toward repetitive action. An ironing cat would also earn both of you a spot on the Letterman show.

These suggestions work best early in the development of the behavior; later, the obsessive-compulsive behavior manifests itself in reaction to almost anything the animal perceives as upsetting, and the cat becomes "hooked" by the behavior and cannot stop. In cases where the cause is not known or cannot be removed, tranquilizers, anticonvulsants, progestins, and other drugs may be helpful.

Table 18:
Signs Associated with
Obsessive-compulsive Disorders in Cats

Self-licking, air licking, hair pulling

Staring into space, imaginary prey chasing, snapping jaws, batting air with paws

Excessive eating and/or drinking, excessive drooling, wool sucking, and eating fabrics

Head bobbing, tail swishing, running, jumping into the air, pacing, freezing

Crying, howling

Chewing of tail or feet, clawing at mouth, unprovoked aggression toward people

PSYCHOGENIC SKIN DISEASE–This is a specific form of obsessive-compulsive disorder that manifests itself as a lick granuloma (red, ulcerated, oozing sore) or as baldness, especially between rear legs and around the tail. In both cases, the cat has caused the skin disease by excessive licking, hair pulling, and biting. This disease is seen more commonly in cats with Siamese, Burmese, Himalayan, and Abyssinian breeding.

Cats that have the bald form of skin disease often have stubbled hair, may show a "stripe" down the back, or may, especially in Siamese cats, have a regrowth of darker hair in affected areas. Your cat's veterinarian makes a diagnosis of psychogenic skin disease based on the history of self-mutilation and negative results on laboratory tests for other skin disorders. Treatment is geared toward topical care of lesions with ointments and/or bandaging and treatment of the psychogenic upset, as described for obsessive-compulsive disorders.

ROLLING SKIN DISEASE–This disease, also called feline hyperesthesia syndrome, is an obsessive-compulsive disorder. My first experience with this interesting syndrome occurred the first week my yellow page ad for Cat Clinic appeared in the telephone book.

A woman called to ask my advice about her Siamese's peculiar behavior during thunderstorms. The cat's behavior as described by her owner went something like this: "When the weather changes, especially when it rains, the cat begins to get a strange look in her eyes; she looks toward her tail and rear as if something is bothering her back here; then she begins to swish her tail; the skin on her back begins to ripple and roll; and finally the cat runs through the house with a wild look in her eyes. If the maid or I happen to get in her way, the cat attacks us. The last time this happened, I had to take the maid to the emergency room because the cat bit and scratched her badly. The cat is gentle and loving except when she has one of these attacks. I just wondered if you were familiar with the problem and had any suggestions."

One of my own personality peculiarities is that I attempt to joke when I am stressed. This was my first week as a cat veterinarian, and I hadn't the foggiest idea about the problem the woman was experiencing with her cat. I said, "Well, we can be grateful for one thing—it rains in Amarillo only twice a year." The woman was not amused. I then suggested that I examine the cat and proceed from there. The woman said she would call me back.

As fate would have it, I discovered the cause of the cat's bizarre behavior that very afternoon. I received my copy of Dr. Jean Holzworth's wonderful book, *Diseases of the Cat,* in the afternoon mail. I unwrapped the book, and it fell open to the page describing rolling skin disease. It is a rippling of the skin of the back; biting or licking at the tail; sudden violent licking when running, walking, or eating; running crazily around the house; and attacking objects, including the owner, without provocation. The cats may even have generalized seizures and assorted problems, such as urinating outside the litter box. Signs may last a few seconds to a few minutes, occur at a specific time of day, vary in incidence from month to month, and occur once every few days or almost constantly all day. Between episodes the cats may be normal or slightly agitated. Once asleep, these cats seem peaceful but may wake up suddenly at night and attack their owners.

The woman had described the exact symptoms of rolling skin disease except for the part about the weather changes. I could hardly wait for the woman to call back so I could show her how knowledgeable I had become about cat diseases, but, of course, she never did. Later, I described the syndrome in a national pet column, and the response from readers indicated that the ailment is not rare.

The treatment is the same as for other obsessive-compulsive disorders, with anticonvulsant drugs being touted as the most successful regimen.

The cause of the disease is not known, but contributing factors quoted in the Holzworth book include "unstable personality, frustration of various kinds, and environmental pressures, with other contributing factors such as toxins, dietary preservatives, or minimal brain damage possibly being involved as well."

Weather change might be considered an environmental pressure. It is well documented in human medicine that individuals vary in sensitivity to weather, and that weather greatly affects physical and mental well-being. Psychotic episodes, depression, and aggressive criminal behavior such as rape, murder, and assault, as well as certain physical disorders such as headaches, occur more frequently during certain weather conditions.

Cats, with their superior senses, must be affected greatly by weather conditions. The cat with rolling skin disease was, I am guessing, extremely sensitive to changes in barometric and atmospheric pressure, and this sensitivity served as the trigger for the mental disorder.

If one knows to expect problems during particular weather changes, precautions such as extra rest, proper nutrition, and alleviation of other stressors might do much to counterbalance weather stresses.

PHOBIAS–*Webster's New World Dictionary* defines a phobia as "an irrational, persistent fear of some particular thing or situation." Although they may vary in degree, phobias are common in cats and include fear of traveling in the car, fear of water, fear of dogs, or fear of certain noises.

Phobias are treated by counterconditioning and by desensitization. We have covered counterconditioning methods in the previous two chapters. To desensitize the cat to a noise phobia, for instance, you would expose the cat to short periods of low-volume noise and reward the cat for nonfearful behavior.

Suppose the cat was afraid of thunderstorms. Buy or make a tape of thunderstorms, and play it very low while distracting the pet with some pleasurable activity or treat. During the next session you would increase the volume or the time the sound was played or both. If the cat becomes fearful, you go back to the previous step.

Question

Dear Dr. Whiteley,

My cat has developed a weird habit that I wish to know about. When the cat was about six months old, she broke her right front leg and wore a cast for several weeks. She recovered fine and showed no limping or other sign of her injury.

When the cat was about a year old, I had her declawed, front feet only. Ever since the declawing surgery, she will limp on the right front leg whenever she walks across a smooth surface—the tile kitchen floor or the brick fireplace hearth, for example. When she walks on carpet or rugs, there is no sign of the limp. I have had the cat examined by the veterinarian who did the surgery, and she cannot find anything wrong. It has been a year since the surgery. What do you suppose is causing my cat to limp?

Robert in Pasadena

Dear Robert,

You may have hit upon the answer when you describe the limping behavior as a habit. I can think of no physical reason for the cat to limp on hard surfaces, so I'm guessing that the reasoning is mental. Suppose that when the cat returned to the hospital for the declawing surgery, she remembered the cold, smooth cage as being the same as when she visited the hospital for the broken leg. Cold, hard surfaces somehow became associated with the broken leg, triggering an unconscious memory to limp. It might be interesting to cover the hard surfaces with rugs and see if the habit disappears after a couple of months.

My apologies to the cat if my guess is wrong.

H.E.W.

10

Diseases Affecting Behavior

Cats are good at hiding signs of sickness. For the most part, they are fastidious in grooming and toilet habits; thus diarrhea may not be apparent. Some cats are relatively sedentary during the day and roam at night; therefore lethargy and inactivity may go unnoticed. Owners may also fail to observe that the cat has reduced his food and water intake if these commodities are readily available.

Yet there are some common signals of physical illness. The sick cat often seeks seclusion and isolation. His grooming behavior decreases in conjunction with his general inactivity, so his hair coat begins to look unkempt. Dehydration and anorexia can contribute to a sunken appearance of the eyes and a loss of elasticity of the skin. The ears and outer extremities may be warm to the touch, giving a clue that the cat has a fever.

Cats, like children, mount a fever response easily when suffering from disease. Actually, the fever serves a purpose: The elevated body temperature increases the body's effectiveness in mounting an immunological defense against infectious diseases.

To help maintain the elevated body temperature, a sick cat will curl up; this reduces the body surface and cuts heat loss by convection and radiation. Heat is also built up by shivering and by fluffing up the hair so that its insulating quality is increased. The cat is lethargic and sleeps more, thus conserving the energy needed to maintain the fever. When an individual mounts a fever response, there is an increase in the body's metabolic rate of 30 to 50 percent.

These are general behavioral symptoms that accompany most illnesses from which the cat may suffer. Specific behavioral changes also accompany specific disease syndromes in the cat.

Rabies

The disease of rabies is well known because of its violent symptoms and fatal nature. The word "rabies" means "rage"; the disease was so named because of the psychic changes it induces in its victims.

For cat owners, the disease is more frightening than ever before because, in recent years, cases of rabies in cats surpassed cases of the disease in dogs. Cats are more susceptible to the rabies virus than dogs; feral cats are more likely than dogs to come in contact with skunks, raccoons, bats, and foxes—the wild animals most likely to carry rabies; and fewer cats are vaccinated against the disease, probably because fewer state and local ordinances require the vaccination of cats. The increase in cases of cat rabies is alarming to public health officials because cats are more likely than dogs to expose humans to the disease.

The clinical course of rabies can be divided into three phases: the prodromal (earliest signs of disease), the excitative (a cat with dilated pupils that attacks suddenly, biting and scratching, or a cat that becomes uncharacteristically friendly), and the paralytic (paralysis of the muscles of the jaw and throat, with salivation and inability to swallow). The paralysis phase progresses rapidly, leading to coma and death in a few hours. A rabid cat will be dead within ten days after the onset of symptoms.

All cats older than three months should be vaccinated against rabies. Booster vaccinations are recommended at one- or three-year intervals, depending on the type of vaccine administered. However, city or state ordinances may require annual vaccination regardless of the vaccine used.

Several years ago I served as a national spokesperson for a campaign to urge cat owners to vaccinate their feline charges against rabies. It was a rewarding job because rabies is such an interesting and gruesome illness, and I picked up interesting tidbits about the disease. Did you know that if you see a fox that has tangled with a porcupine, evidenced by a quill sticking out here or there from his anatomy, there is over a 90 percent chance that he is rabid? A fox in his right mind is too smart to confront a porcupine. If you see a dog with porcupine quills it means nothing, for dogs are not that smart (at least where porcupines are concerned), and veterinarians extract porcupine quills from the same canine victims over and over again. Alas, my research revealed nothing about the porcupine IQ test for cats.

Food Hypersensitivity

Although not as dramatic a problem as rabies, food allergies account for a significant number of cases of itchy cats that scratch and bite at their skin and hair. While this behavior has a disease origin, it can become an obsessive-compulsive disorder if left untreated.

The goals of therapy are treatment of skin and hair complications while changing the cat to a diet free of allergy-causing ingredients. The hard part of this regime is determining the allergy-causing ingredients. This can be accomplished by blood or skin tests, which hopefully reveal substances to which the cat is allergic, and by placing the cat on a hypoallergic diet, either prescription or homemade, free of ingredients causing the allergic reaction.

Benzoic acid, a food additive known to produce feline skin allergies, has also been proven to play a role in aggressive misbehavior and hallucinations in cats.

Poisoning

Cats differ significantly from most mammals in their metabolism of drugs. One extra strength (500 mg) or two regular strength (325 mg) acetaminophen (Tylenol) capsules given four hours apart can produce signs of anorexia, vomiting, and depression within one to two hours of administration. Products containing acetylsalicylic acid (aspirin) must also be given with caution, as this drug stays in the cat's body for a prolonged time when compared with dogs or humans. Even such commonly used products as hexachlorophene (pHisoHex) and Pepto-Bismol must be used with caution. Consult your cat's veterinarian before administering over-the-counter drugs or medications intended for dogs to your cat.

Parasiticides and insecticides containing organophosphates in high doses can cause excessive salivation, convulsions, and death in cats. Organophosphate poisoning occurs commonly when cats are dipped or sprayed with products designed for flea and tick control in dogs or large animals. Lindane and other cholinated hydrocarbons commonly used as insecticides should never be used on cats. Always use products labeled for cats according to directions.

Marijuana is as enticing to a cat as catnip. Symptoms of marijuana intoxication in the cat are staring at the wall or floor, increased reaction to noise, and twitching of the skin.

Lead accounts for a significant number of cases of accidental poisoning of cats. Lead toxicity occurs more often in summer and fall because of increased exposure to outdoor sources of lead, which include lead-based paint, solder, putty, hard water from lead pipes, and batteries. Cats are less likely than dogs to ingest lead directly, but absorb toxic amounts by removing lead-containing substances from their hair coat by grooming. Vomiting and diarrhea accompany behavioral changes, which include hysteria, blindness, convulsions, aggression, and head pressing. Diagnosis is confirmed by laboratory tests, and treatment is aimed at preventing further exposure, decreasing absorption from the gastrointestinal tract, and providing supportive care.

Ethylene glycol (antifreeze) poisoning occurs much too frequently in fall, winter, and early spring during peak usage. This sweet-tasting substance is lethal in small amounts: A teaspoonful will kill a seven-pound cat. Common symptoms include vomiting and diarrhea, depression progressing to seizures, coma, and death. Early diagnosis and treatment are imperative if the poisoned animal is to survive.

Many preparations used to kill rodents contain an anticoagulant, leading to hemorrhage in animals that consume them. Cats become poisoned directly by eating the rodent bait or secondarily by consuming rodents that have consumed the bait. Signs of poisoning are those associated with blood loss—bleeding from body cavities, pallor of mucous membranes, depression. If bleeding occurs within the brain or spinal cord, nervous system signs and behavioral changes will be seen. Treatment for cats diagnosed with this toxicity is vitamin K administered by injection and/or orally.

If you suspect that your cat has been poisoned, contact your pet's veterinarian and the National Animal Poison Control Center (NAPCC), a nonprofit, animal-oriented poison center located at the University of Illinois. This center is staffed twenty-four hours by veterinary health professionals, and they are more familiar with poisons, symptoms, and treatment of intoxication in small animals. A charge is made on your VISA, MasterCard, or American Express card at the rate of $2.95 per minute if you call the 900 number and $30 a case if you call the 800 number. If the product suspected of causing poisoning is covered by a sponsoring company, the call is free.

Table 19: Contacting NAPCC About a Poisoning Case

Call 1-900-680-0000 or 1-800-548-2423.
Have charge card information available.
Provide your name, address, and telephone number.
List species, breed, age, sex, weight, and number of animals involved.
Provide the name of the poison your cat or cats have been exposed to, if known.
Give information concerning the poisoning—amount of the poison, the time since exposure.
Provide a summary of any problems your animal is experiencing.

Table 20: Houseplants Potentially Toxic to Cats

Amaryllis	Hyacinth
Azalea	Hydrangea
Bird of paradise	Jack-in-the-pulpit
Cacti	Lily of the valley
Caladium	Mock orange
Calla lily	Monkey pod
Daffodil	Philodendron
Daphne	Poinsettia
Dumb cane	Rhododendron
Elephant's ear	Snow-on-the-mountain
English ivy	Yew
Holly	

Thyroid Disease

The thyroid gland, located in the neck, produces hormones that regulate many of the body's functions. Tumors of this gland, usually benign and occurring in older cats, cause an increase in circulating thyroid hormones.

The cat suffering from an increase in thyroid hormones (hyperthyroidism) shows several behavioral signs: He becomes hyperactive, experiences sleeplessness, paces the floor, and develops a voracious appetite. Physically, however, the cat loses weight, and his heart beats rapidly. You can often put your hand over the chest of a cat suffering from hyperthyroidism and feel the runaway heart.

Diagnosis of this disease is made by blood tests revealing an elevation in thyroid hormones. Treatment is aimed at reducing thyroid hormone levels by use of oral antithyroid drugs, removal of the thyroid gland by surgery, and treatment with radioactive iodine. The last two therapies are best performed by veterinary experts: The surgery is difficult, and radioactive iodine is limited to universities or private practitioners approved to use medical radiation.

I remember two cats in particular that suffered from this disease. The first cat, one of my favorite patients, was elderly and belonged to a personable single woman in Wisconsin. The cat showed every symptom of the disease described here, but this was years ago before hyperthyroidism or cat diseases in general were receiving much attention in the veterinary press. A year after the cat died I attended a lecture by Dr. Barbara Stein, an authority on feline medicine and surgery, and discovered the diagnosis. Alas, I remember all of my known failures vividly.

The second cat, also elderly, which I treated last year, was one of the meanest and most difficult of all my patients; she belonged to a congenial woman and her elderly mother, and both women doted on this cat, plus seven other cats. It was hard to pinpoint a change in condition or in activity level with this cat, for she had been thin and hyperactive her entire life; the clue, however, was that the cat had begun pacing all night, and when I finally got close enough with my stethoscope for a listen, the cat's old ticker was racing for all it was worth. With great difficulty I obtained the necessary blood for laboratory tests, and with great reluctance I told the cat's owners the resulting diagnosis. The cat was impossible to treat with pills or any other medication, so the owners decided to put the cat to sleep.

When the thyroid gland produces too little of the thyroid hormones, the cat's body is also prone to upset. Traditionally, the hypothyroid person or animal was thought to be overtired, sluggish, obese, and susceptible to colds. In the early seventies, this was the popular diagnosis in the human medical field, and I, like millions of other women, took my daily quota of thyroid pills. Only years later, when I tapered myself off the drug and submitted my own blood for analysis, did I discover that my thyroid gland was working just fine. My sluggishness, tiredness, and chubbiness had other causes, mainly demanding jobs, a husband and children, and an overfondness for sweets.

Hypothyroidism does exist, both for women and for cats, but it doesn't always manifest itself in the typical symptoms of underactivity and obesity. Recently, several studies in canine medicine have shown a behavioral syndrome associated with a reduced level of thyroid hormones; symptoms include aggression and extreme shyness and seizures. It is almost certain that a similar syndrome exists in cats.

Epilepsy

Epilepsy is a condition of recurring seizures. An animal undergoing a seizure loses control of his muscles and limbs, may salivate, and may show behavioral changes.

Rather than a disease, epilepsy should be considered a symptom or sign of disease because underlying causes can be as varied as brain tumor, stroke, trauma, poisons, liver and kidney damage, thiamine deficiency (vitamin B_1), congenital disease such as hydrocephalus (water on the brain), and inflammatory diseases.

Inflammatory diseases affecting the brain can be caused by viruses (feline leukemia virus, feline immunodeficiency virus, feline infectious peritonitis), bacteria (bacterial meningitis), or parasites (toxoplasmosis, *Cuterebra* larvae).

Diagnosis of epilepsy is never easy. If, however, a causative condition is discovered, treatment is aimed first at the underlying disorder. For example, if thiamine deficiency is a factor, the cat is placed on a well-balanced diet. If the cat tests positive for feline leukemia virus, drugs such as interferon which enhance the immune system may be tried. If a brain tumor is diagnosed, surgery may or may not be an option.

The seizures that manifest themselves as a primary or secondary symptom are controlled by anticonvulsant drugs.

Toxoplasmosis

Most cats that become infected with this protozoan parasite show no sign of disease. The only indication that an infection has occurred might be the presence of oocysts (eggs that contain encysted organisms) on fecal examination or an elevated antibody titer on blood testing. When the disease does occur, usually in young or immunosuppressed cats, it manifests itself most often as bloody or mucus-containing diarrhea or as pneumonia. Occasionally, however, cats suffering from this parasite will show damage to the eyes or brain.

Cats with eye or brain damage exhibit behavioral signs, governed by the part of the brain affected. The symptoms may include convulsions, sleepiness, head pressing, grinding of teeth, personality changes such as excessive aggression or friendliness, somersaulting, circling, and trembling.

Cats diagnosed early are treated with sulfa drugs and with supportive therapy; cats showing signs of severe illness rarely survive.

Cats acquire the parasite either by ingesting sporulated oocysts from the feces of other cats, which can remain infective in soil or sand for long periods in some climates, or by ingesting wild animals or uncooked meat contaminated with the parasite. Humans can also acquire the disease by eating undercooked meat contaminated with the parasite, and by ingesting oocysts contained in an infected cat's feces or in contaminated soil or on contaminated cooking surfaces.

Because *Toxoplasma* organisms can cause birth defects in children born to infected women, this disease has often spread panic among pregnant cat owners. In several unfortunate cases, a pregnant woman's physician has urged her to get rid of the family cat, and she has complied.

I believe that the panic is unwarranted. Approximately half the people in the United States, including 25 to 45 percent of the women of child-bearing age, are already immune to the disease because they have been exposed to the parasite and have developed antibody protection without ever experiencing symptoms. For most adult humans, symptoms, when they do occur, mimic the flu. The human population endangered by the disease are people with suppressed immune systems such as AIDS patients and pregnant women with no previous exposure to the disease who become infected shortly after conception; even among these women, there is less than a 30 percent chance that the fetus is harmed. However, no risk is tolerable, and I always advise pregnant or immunosuppressed clients to take the following precautions:

Table 21: Human Protection Against *Toxoplasma*

Cook all meats to a minimum internal temperature of 151 degrees Fahrenheit, and carefully wash your hands and all contaminated surfaces after handling raw meat.

Prevent your cats from hunting wild animals and birds, and do not feed them raw or undercooked meat.

Wear gloves or have another family member take over litter box duties. Cat feces should be flushed down the toilet or burned within twenty-four hours. Litter boxes should be cleaned with boiling water daily (*Toxoplasma* oocysts do not become infective until one to five days after being passed in the feces).

Wear gloves while gardening in areas frequented by cats. Children's sandboxes should be covered when not in use.

Good personal hygiene after animal contact is important. It is also important to control flies and cockroaches, which can transport infective eggs to food.

For additional peace of mind, you can have your cats and yourself tested for the presence of antibodies to toxoplasmosis. If both of you have a significant level of antibodies against this parasite in your blood, the likelihood of either becoming infected is slight. If neither of you has antibodies, you are both susceptible to infection if exposed.

Feline Leukemia

Hal, short for Halloween, is a young orange cat, and one of my favorite patients. I first saw him at midnight when he was three months old, after he had gotten into a serious argument with a big dog. Two weeks later Hal's "mom" called to tell me that Hal had recovered from his fight wounds but that, unbeknownst to her, Hal had gone to school with her to pick up the kids. Hal had ridden on top of the van and must have put on a great balancing act to make it to school intact. "I wondered," Hal's owner said, "why all these cars kept honking at me." Not long after, Hal had another midnight mishap, this time with the neighborhood bully cat. He recovered rapidly, and I was convinced that he lived a charmed life.

When Hal was six months old I neutered him and drew a blood sample to check him for feline leukemia virus (FeLV) prior to vaccinating him against the disease. The test was positive, and I wondered if the little fellow's luck had run out.

The virus that causes feline leukemia is transmitted from cat to cat through body secretions, saliva primarily. Because the virus can survive outside the cat's body only a short period, close contact with a cat shedding the virus is necessary for the spread of the disease. Approximately 20 percent of kittens from mothers infected with the virus will be born with the infection.

One-third of all cancer deaths in cats are FeLV related, and greater numbers of cats die from anemia and infectious diseases caused by the detrimental effects of the virus on the cat's immune system. When the reproductive system is infected, infertility, abortions, fetal resorption, and birth of weak and fading kittens may result. When the neurological system is affected, cats show signs such as paralysis, depression, and convulsions. Whenever I examine a cat with multiple unusual symptoms, I always check for feline leukemia virus.

Feline leukemia is one of the diseases for which preventative vaccine is available. See Table 22, page 189. Many veterinarians test the cat for the virus prior to vaccinating him. The test is performed on a sample of blood, saliva, or tears, depending on the type of test selected. I recently noticed advertising for at-home FeLV test kits. These kits are designed for the owner to take samples of saliva or tears, which are sent to a laboratory. I recommend that you use the services of a veterinarian for FeLV testing because interpretation of test results and decisions about treatment of your cat if he tests positive are critical.

When Hal tested positive for feline leukemia, he appeared to be in perfect health. He lived in a house with one older cat and two littermates. The older cat had been vaccinated against feline leukemia, but I retested him and the siblings. All were negative. I vaccinated Hal's sisters and told Hal's anxious mistress that a positive FeLV test was not the kiss of death, especially for a cat with Hal's will to live. I treated Hal with the immune-enhancing drug interferon and retested him for FeLV in six weeks. He was negative at that time, and I vaccinated him against the disease. Hal has had no further problems with FeLV, but he recently caught his foot in a rusty tin can and required twenty sutures.

Hal, like all young cats, is highly susceptible to infection with FeLV. If cats become infected, several things can happen. In Hal's case, his body's immune system came to the rescue and enabled him to overcome the infection. In some cases, the body mounts only a partial immunity; the cat will appear normal and even test negative, but the cat's body hides little pockets of the virus. If this cat is stressed severely by pregnancy or treatment with drugs such as cortisone, he may develop an extensive FeLV infection and become ill. Other cats become infected at an early age, appear normal, but shed the virus in body secretions, thus becoming a hazard to other cats. Most of these cats die of FeLV-related illnesses within three years.

Advice must be tailored to individual situations. If all cats that test positive for FeLV have been removed from your house, you can safely bring a new cat home after a waiting period of a couple of weeks. If you have a multiple-cat household and all of your cats have tested negative and been vaccinated against FeLV, test and vaccinate new cats before allowing them to become part of your family. If you own a cat that tests positive for FeLV and appears healthy, I advise retesting at six- to eight-week intervals. To be on the safe side, clean or replace litter boxes and food and water dishes shared with other cats; isolate this cat from your negative cats. Decisions about the care or disposition of sick cats with FeLV or healthy-appearing cats that repeatedly test positive for FeLV should be made with a veterinarian who is familiar with your cats and situation.

Feline Immunodeficiency Virus

A virus named feline immunodeficiency virus (FIV), causing an AIDS-like disease in cats, was recently discovered. In spite of the similarities between HIV (human immunodeficiency virus), which causes human AIDS, and FIV, people cannot get AIDS from cats, and cats do not get FIV from people.

FIV is believed to be transmitted through saliva, with the most likely transmission being through bite wounds. Free-roaming male cats between five and seven years of age are statistically at greatest risk for acquiring FIV infection. Cats may become infected by sharing contaminated food and water utensils, and by being groomed by an infected cat. Unlike human AIDS, sexual transmission is not a major source of infection in cats.

The clinical signs of cats infected with FIV are similar to those of people suffering from AIDS and of cats infected with feline leukemia virus. These diseases weaken an infected individual's immune system, making him more susceptible to bacteria, viruses, and parasites. When FIV directly infects the brain, personality changes, psychotic behavior, facial twitching, and seizures may be seen. Cats experiencing nonhealing sores, persistent infections of mouth and gums, anemia, and chronic diarrhea should be checked for FIV by a blood test that detects antibodies to the virus.

Although it appears that an infected cat remains infected throughout his life, months or years may pass before the cat suffers from a life-threatening illness. Treatment depends upon the presenting symptoms, as there is no specific cure. The infected cat should be isolated from uninfected cats.

Feline Infectious Peritonitis

Feline infectious peritonitis (FIP) is another virus for which there is no effective treatment. FIP is an immune-mediated disease, meaning that the cat's immune system goes awry in response to the viral infection and causes symptoms of disease.

Symptoms take two forms—accumulations of fluid within the abdomen or chest (wet form) or production of granulomatous tissue in various parts of the body (dry form). Symptoms depend upon what part of the body is affected. If the brain or spinal cord is infected, personality changes, convulsions, and unsteady gait will be seen. If the reproductive system is infected, reproductive failures and "fading kitten" syndrome will be seen. If the eye is infected, cloudiness and spots appearing on the visible surface of the eye may be seen. As the disease progresses, persistent fever, anorexia, and depression develop. Death is the most common outcome.

Laboratory tests are helpful to making a diagnosis but are not as specific as those for FIV or FeLV. The virus, shed in the feces or other secretions of cats immediately after exposure, is not infective to contaminated surfaces such as dishes and blankets longer than two to three weeks. Preventive vaccines are now available.

It is because physical diseases can manifest themselves as personality or mental disorders that all animals showing behavioral changes should receive a proper physical examination, including laboratory tests, before the diagnosis of misbehavior is made.

Table 22:
Feline Vaccination Schedule

Your veterinarian's recommendations may vary from the schedule below, as he or she advises you of a vaccine timetable based upon your cat's individual and cattery history.

Disease	Initial Vaccination (age in weeks)	First Booster	Additional Boosters
Panleukopenia (distemper)	9	3–4 weeks later	Annually
Rhinotracheitis (respiratory disease)	9	3–4 weeks later	Annually
Calicivirus (respiratory disease)	9	3–4 weeks later	Annually
Chlamydia psittaci (respiratory disease)	9	3–4 weeks later	Annually
Feline leukemia	9	3–4 weeks later	Annually
FIP (Feline Infectious Peritonitis)	16	3–4 weeks later	Annually
Rabies	12	1 year later	Annually or Triennially

Most vaccines protect against multiple diseases, thereby reducing the number of inoculations required.

Questions

Dear Dr. Whiteley,

My cat Bozo never goes outside and is normally a calm and gentle cat. However, he recently bit the neighbor's child who was here visiting one afternoon. The next day the local animal control people called and said that I have to have Bozo boarded at a veterinary hospital or at the pound to prove that he doesn't have rabies.

I am upset and mad. I can't bear to think of my baby in jail, and I think this is just another way that the city bureaucrats have to harass good law-abiding citizens. Don't you agree?

Aggravated in Oklahoma City

Dear Aggravated,

I'm sorry, but I don't agree. The animal control people are following what is considered standard procedure for dealing with animal bites to people. The ten-day quarantine at the animal shelter or veterinary hospital ensures that the animal will not escape and that, if he is living at the end of the quarantine period, he could not have transmitted rabies to the child.

I realize that the chance of Bozo having rabies is low, but since the disease is fatal to people and animals the safety of the exposed person is always the primary consideration. I might add that in the early eighties a young girl in Michigan, a state with a low incidence of rabies in the wildlife population, contracted rabies in her own home from a bat. Bats and raccoons can gain entrance to houses, and cats that "never go outside" have been known to sneak out at times.

Curb your anger and go along with the quarantine in the spirit of neighborly love.

My best!

H.E.W.

Dear Dr. Whiteley,

My cat, aged two, was recently diagnosed as having feline leukemia. The cat is ill, and I am treating her at home. My neighbor, who is a busybody, has been over every day since I brought the cat home from the veterinary hospital. Convinced that the cat is going to give my two children leukemia, she comes over to lecture me about putting Buffy to sleep. My veterinarian has assured me that the disease cannot be transmitted to people, but I'd like another opinion.

Worried in New York

Dear Worried,

You have enough stress caring for children and a sick cat without your neighbor coming over to give you a guilt trip.

There is no evidence that the virus that causes feline leukemia can be transmitted to people. I have served as my own guinea pig on the subject. When I was teaching a laboratory course to veterinary technology students, I accidentally swallowed an entire pipette full of serum that carried the feline leukemia virus. And here I am seven years later, fat and sassy and clumsy as ever around laboratory equipment.

If and when the time comes to put the cat to sleep, I'm sure you will make the right decision for all concerned.

Good luck!

H.E.W.

11

Sex

Cats, like most mammals, have their young in the spring when plants and prey are abundant, and the weather is mild. Female cats (queens) are pregnant a short time—approximately nine weeks—and, therefore, begin mating when daylight hours are increasing at the close of winter and the beginning of spring.

An increase in daylight hours signals the queen's brain to produce hormones that activate the reproductive system. A cat begins to experience heat cycles usually in February, one to two months after the winter solstice, and these heat cycles continue until past the summer solstice, ending in September or October.

Artificial light of sufficient magnitude maintained for twelve to fourteen hours a day will bring females into heat during November, December, and January when most cats take a breather from sex. It takes a light source of greater intensity than what is available in most homes, however, to fool Mother Nature and a cat bent on her time off.

The Female

Schematic Female Reproductive Tract

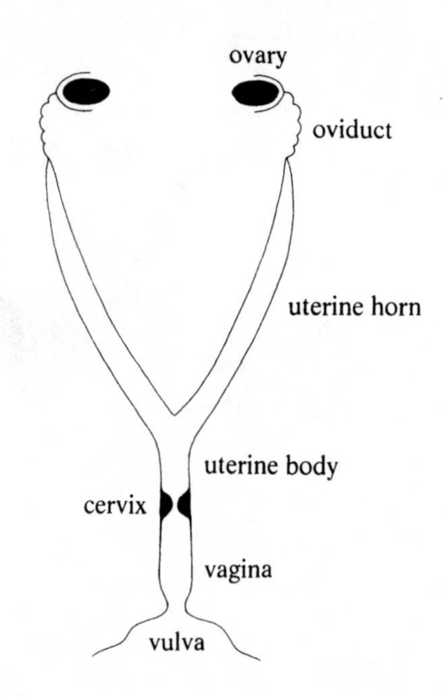

The female feline genitalia consist of two ovaries, each connected to a uterine horn by oviducts, or Fallopian tubes. The uterine horns converge at a single small uterine body closed by the muscular cervix. The cervix, which dilates during labor, joins the vagina, which becomes the vulva, the external genitalia visible to the eye. It is in each of the uterine horns that kittens develop in the pregnant queen.

The genital arrangement of the female dog is essentially the same. The opossum, however, has an interesting reproductive tract arrangement: The female possesses two separate vaginas, uteri, and cervixes. The male opossum, in order to fulfill his lovemaking duties, has a forked penis. Sorry, I couldn't resist inserting this bit of trivia; it might appear on "Jeopardy."

The ovaries of the cat become active and begin to secrete estrogen, the hormone that induces the heat cycle, when the cat reaches puberty between six and nine months of age, although some individuals may begin cycling as early as four months and others as late as one year. Purebreds usually reach sexual maturity more slowly than mixed breeds. Close proximity to a cat in heat will often activate an associating female's heat cycle.

Females younger than one year and older than seven years often have irregular heat cycles, and older feline mothers tend to have smaller litters and more reproductive problems than their younger sisters. Eighteen months to seven years is considered the optimum age for breeding female cats.

The female cat's heat and breeding behaviors are uniquely designed to perpetuate the species; these seductive postures and come-hither vocalizations have inspired the "cat" words that populate our street sexual language—catting around, cat house, and others I'll refrain from mentioning.

Owners who are unfamiliar with the female's heat behavior often call me in a panic about their cat's crazy behavior. Heat behavior begins with an increase in the cat's activity and nervousness, and may coincide with a reduction in food intake. The cat begins to sing a mating song, sounding much like the lead singer of a modern hard-rock music group, hoping to attract a mate with her lyrics.

The female in heat may also spray urine, which contains odorous clues that the cat is receptive to boyfriends. Urine can elicit the Flehmen (lip curl) response in interested males up to three days after being sprayed on a tree or other object.

The cat in heat rubs her head and flank against objects and people in order to spread her perfume around her territory. She may roll over and over on the floor, sometimes gently and sometimes violently. Her seductive behavior causes male cats to voice their own serenades of love while loitering on the street corner, on the doorstep, or in the backyard of the female's territory.

The external genitalia of the female cat are small enough that its swollen appearance during heat is hard to detect; there is no bloody discharge from the vulva, as in the dog, to give evidence that the cat is in heat. The clues to the cat's amorous intentions are primarily cytological and behavioral.

Cytology is the study of cells, and the cells lining the vulva of the female change in accordance with the stage of the heat cycle. It is possible for your cat's veterinarian to take a smear of the cells lining the external genitalia of the cat and determine under the microscope if she is in standing heat, termed estrus, and ready for breeding.

At the beginning of the heat cycle, at a stage called proestrus, males are attracted to the female, but she is not receptive to masculine advances; as her hormones progress her into true heat, estrus, she invites males and stands to receive one or more of them. It is at this stage that breeding and conception take place.

During heat the cat orients herself posturally toward a partner, and that partner is occasionally human. It is possible for young cats to be sexually imprinted upon their owners. The cat looks at the partner, feline or human, and turns her vulva toward him. She assumes a receptive position by elevating her hips and lowering her back; she holds her tail to one side so that it will be out of the way. The cat squats on her back legs and may tread back and forth with her rear feet. This behavior is reflexive: Cats have been seducing males by these actions since the first cat seduced "Tom" in the Garden of Eden, and the behavior can be induced by pressing lightly in the middle of her back or by grasping the skin over the nape while stroking her rump.

The average duration of estrus behavior is a week, but may

last anywhere from one day to three weeks with an interval of approximately eight days before it starts all over again in the non-pregnant female. Many canine caretakers are shocked to find that their new feline companion does not follow a heat cycle of every six months like a dog. "Do you mean," they cry, "that I have to put up with this yowling and increased activity every week or so?"

"Yes," I reply. And so springtime, that time of wonderful awakening of sexual activity in the feline population, initiates the bulk of feline owners to schedule their amorous kitties for spaying (ovario-hysterectomy) surgery. If the owner decides to let nature take its course and allows the cat access to a tom, kittens usually follow nine weeks after the female's loud and sensuous heat behavior.

Table 23:
Heat (Estrus) Behavior Timetable

Heat behavior commences at puberty—six to eight months of age average.

Puberty may be activated early by close proximity to breeding toms and/or other females in heat. Puberty often delayed in purebreds raised as solitary pets.

Heat cycle activated by increasing daylight hours; ceases in response to decreasing daylight—usual breeding season from late January to October.

Heat cycles occur irregularly in intact females older than seven years and in youngsters less than a year old.

Average heat duration is seven days.

Period between heat behavior approximately ten to fourteen days.

Heat commences three to four weeks following weaning of previous litter.

Heat commences within a week of abortion of kittens.

The Male

Schematic Male Genitalia

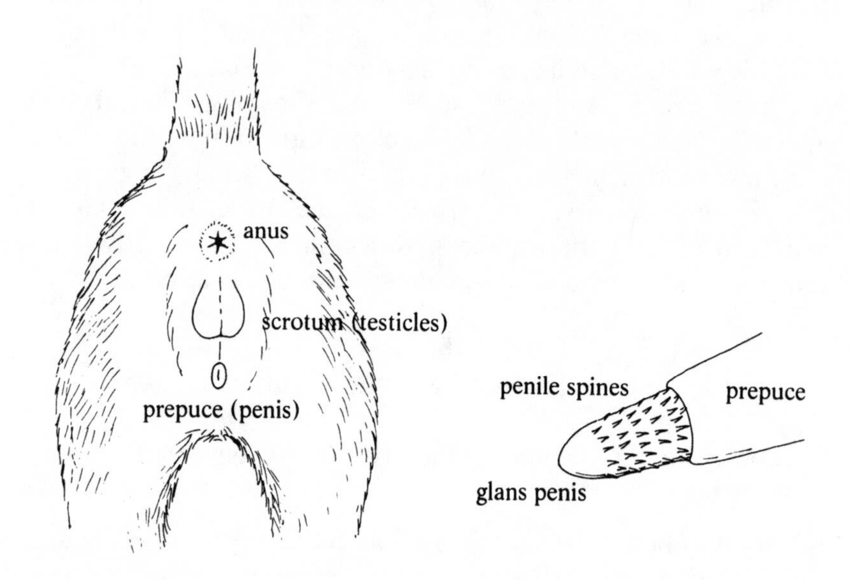

The genital arrangement of the male cat is the opposite of the male human. The penis is located below the scrotum and directed backward. A major portion of the feline penis is covered with spines, approximately one hundred to two hundred raspy projections, facing the body of the cat.

The male kitten's testicles are descended into the scrotum at birth, and by three and a half months of age the testicles are developed enough to produce testosterone, the male hormone. Testosterone initiates the growth of penile spines, which reach mature size when the kitten is six to seven months old. The spines begin to disappear two to four months after neutering because the source of testosterone has been removed. At seven months the male kitten's testicles are capable of producing the sperm that will impregnate a female.

Although young males begin sexual play—mounting, pelvic thrusts, and gripping the neck of a play partner—at around four months, they are not usually able to complete intercourse by inserting the penis into the female's vagina (intromission) until after seven months. The healthy adult tom is capable of mating throughout his life.

As the male testicles develop and produce testosterone, heralding the onset of puberty, certain unwanted male-related behaviors may manifest themselves—roaming, aggression toward other males, and urine spraying. Spraying becomes particularly objectionable as the tom grows older because his urine contains strong-smelling chemical substances called pheromones, used to mark territory and attract females in heat. Testosterone also affects the male's physical appearance: An intact male's head becomes large in proportion to his body; his jowls are thick and meaty; and his body is muscular and angular, in contrast to the female, which develops rounder contours because of an increase in fat deposition beginning at puberty.

During breeding season—February through September—the outdoor male's territory and range size increase, depending upon female density and distribution. A male dominant in the brotherhood of breeding toms will have a range larger than that of subordinate males.

The indoor tom is also aware and protective of territory. When he is presented with a female in heat and new surroundings, he is torn psychologically between establishing his right to the new territory and sex. If making feline babies is the intention of the new liaison, the male is more comfortable and less confused about his sexual duties if the female in heat is brought to his house or pen, rather than the other way around.

The male is capable of breeding a receptive female throughout the year, but he is occasionally less eager, most notably during the fall. Luckily, this coincides with the time that females are not experiencing heat.

Cats, like most mammalian species, are polygamous—they mate with many partners. Although exceptions occur, the longest that most feline males or females remain loyal to a sexual partner is one heat period.

Mark Walters, in his book *The Dance of Life: Courtship in the Animal Kingdom,* writes his theory about why polygamy makes sense: "Domination of parenting by the female, symbolized by the exclusive claim to the breast, often makes polygamy a worthwhile alternative for male mammals—worthwhile, that is, in the sense that this route will allow for the survival of the maximum number of his offspring. It is little wonder that almost all birds are monogamous, but that ninety-seven percent of all mammals are polygamous. Where the opportunity for meaningful male participation in rearing young is available, one finds a different story."

Some scientists also suggest that the polygamous lifestyle promotes a disparity in body size and shape between males and females. In polygamous animals, males are larger than females because of breeding competition between males. In monogamous species (those that mate for life), males and females tend to be more similar in size and body shape.

Some owners, capitalizing on the cat's polygamous nature, wish to train their little tom to be a stud cat. Providing a normal kittenhood is the first prerequisite. As discussed in Chapter 2, males raised without adequate interactions with their own species often grow up to be inadequate lovers. The poor fellows just don't seem to know what to do. The next best thing to a great kittenhood is an experienced teacher. When the tom is about a year old, he should be introduced to a very receptive, preferably experienced, female. The love nest should be an area where the male is comfortable and one in which he begins to associate with the act of lovemaking. If he rejects the advances of the selected female in heat, another might be introduced. Some cats are selective in their choice of a mate.

The Act

Cat lovemaking usually occurs at night, and although most people are familiar with the loud and amorous noise that goes along with the breeding act, many have missed out on what really goes on.

The tom calls to the female in a loud voice, called caterwauling. Caterwauling serves two purposes: It advertises the male's availability, and warns other males of his intentions and territory.

The tom's interest in the female increases when he is summoned by her own calls of love or smells her receptivity from urine or secretions from the vulva. Eventually, the male must make his advance: He greets his mate by rubbing his head or nose on hers or by sniffing her genitals. The expression on the male's face as he smells the female in heat is called the Flehmen (the female's sex pheromones activate the secondary scent organ in the roof of his mouth). He wrinkles up his nose and appears to be grinning.

The male then emits a softer cry that indicates his readiness to mate and begins to circle her. If the female is not yet receptive, as sometimes occurs during proestrus, the period before true estrus or heat, she may engage in a type of flirting chase, refusing to let him mount. If she is receptive, she usually assumes the mating stance—pelvis elevated, tail held to side, back legs crouched low to the ground.

The male than grasps the neck of the female with his teeth, using his grip to orient his mount, not to inflict injury. The male swings on top of the female, straddling her first with his forelimbs and then with his hind limbs. He begins treading with his back legs while sliding his pelvis backward. His penis becomes hard as he arches his back and begins hunching movements to introduce his penis into the female's vulva. During ejaculation semen is deposited in the vagina near the cervix. The tom remains motionless for a few moments after intromission; during this time the female becomes excited and her eyes dilate. After the male ejaculates she pulls away, strikes out at him with a talon-exposed paw, and screams.

One explanation for the postcoital scream by the female is that the spines on the withdrawing male penis hurt. Of course, no one but the queen knows, and she won't talk. The penile spines are also believed to provide the stimulation necessary for the female's ovaries to release their eggs.

The domestic cat is an induced ovulator, meaning that eggs are released from the ovaries in response to copulation. There is a peak in the hormones causing the eggs to be released from the ovaries approximately four hours following multiple matings. Rarely—less than 50 percent of the time—does one copulation bring about the release of eggs. It usually takes four matings to result in ovulation and pregnancy, and it is not unusual for a female to mate ten or twelve times in rapid succession.

Let's go back to the act. The entire process of greeting, mounting, and ejaculation takes from one to four minutes. After the male slides off the female, he retreats from her offensive swat and sits nonchalantly nearby, licking his penis and forepaws. The female licks her vulva and then engages in an after-reaction of rolling and rubbing on the floor.

The male has a period of five to fifteen minutes before he is capable of mating again. As the number of matings increase for the hardworking fellow, the refractory period increases. Even the most macho of individuals has a limit. The female may be raring to go before the male she has initially mated with is ready. She may court him unabashedly, moving her rear toward him until she is waving it in his nose, rubbing her head against his, and standing in front of him, kneading and purring.

Other toms may be standing by, waiting their turn. A fight may ensue to determine who will be next. The female usually chooses and may mate with several different males, resulting in a litter fathered by different toms. The female may also have a favorite male, one for whom she waits for her multiple matings.

Not all cats run a "cat house" act. If a female is shy, she may be intimidated by the whole business. In that case, bringing her into the male's home several days in advance of breeding will allow her to become acclimated to the male's surroundings and smell before he is introduced to her. Using an experienced male is helpful. If she still refuses his advances, bring in a second choice.

Cats, by the way, rarely participate in what might be considered sexual rape. For one thing, it is difficult for the tom to achieve intromission if the female does not elevate her pelvis and move her tail aside. For another, rape is considered an aggressive act of control and power, and cats engaging in mating behavior are merely interested in sex.

Exceptions to normal sexual behavior increase when cats and other animals are placed under stressful conditions. Rats, which normally adhere to strict mating protocol, resort to rape and gang rape when forced to live in overcrowded conditions. Murder and infanticide in animals also increase when overpopulation becomes a problem. Catnip, which acts upon some cats like a drug, may also serve to increase sexual aggression in cats of both sexes.

Sexual Problems (Male)

Objectionable Mounting

As discussed earlier, a dominant male, intact or castrated, may become a despot in his own household; he shows his dominance by mounting other cats, regardless of sex. Although this might be described as bisexual or homosexual behavior, it is not indicative of sexual preference for males. When given a choice, the intact male will select a female in heat.

Castration is recommended for the tom indulging in this misbehavior. Remote punishment with water sprayer, administered when the cat is indulging in the misbehavior and in such a way that the cat associates the punishment with the behavior and not with you, may also be helpful.

Castrated males who were experienced in the sex act before neutering may continue the mating postures of mounting and neck grasping of females after surgery. It is estimated that 10 percent of castrated cats indulge in tomcat behavior. In these cases, the learned component of the sexual act is retained, while the source of hormones that govern male behavior and the production of sperm is removed. The male cannot impregnate the female, and his interest in sex wanes but not completely (mind over matter). Oral or injectable progestins usually alleviate this residual interest in sex.

Masturbation might be classified as objectionable mounting. It occurs less frequently in cats than in dogs and is a problem primarily in young isolated male kittens or in young kittens housed in pairs.

Hair Rings

Males that successfully perform neck grip, mounting, and penile thrusting behavior but fail to achieve penetration may be suffering from hair rings. The hair ring is just what it sounds like—a ring of hair accumulating around the base of the penis and held in place by penile spines. The hairs that make up the ring come from the male's prepuce or from the hairs of the female when the erect penis was rubbed over the fur. The hair ring can be removed by gently sliding over the edge of the penis; in some cases, the tom successfully removes the hair ring with his teeth.

Cryptorchidism (Retained Testicles)

As mentioned earlier, a male kitten's testicles have usually descended from the pelvic cavity into the scrotum at birth. If one or both testicles have not descended by six months of age, they rarely do so on their own. If a cat has one descended testicle, he can mate and produce offspring. If both testicles remain within the pelvic cavity, the cat is sterile.

Because of the increased risk of cancer in retained testicles and because this is an inherited disorder, these animals should be neutered. The surgery is more time consuming than normal, for the veterinarian must open the pelvic and/or abdominal cavity to search for the missing testicles.

Infertility in Males

Hair rings and retained testicles are not the only causes of male sterility. Blue-eyed white cats are prone to lowered fertility because of inherited defects affecting hearing, sight, and the immune system as well.

The tortoiseshell- or calico-colored male is a rare fellow because the color genes for orange, black, and white, which make up the calico and tortoiseshell colorations, are located on the female sex chromosomes. The male, representing less than 2 percent of calico and tortoiseshell cats, is sterile; he has an extra female chromosome. Basically, a female has the XX arrangement of sex chromosomes, the male has the XY configuration, and the male calico and tortoiseshell are an XXY.

Is the castrating female a myth? Apparently not in the cat world. A dominant female may so traumatize a young male during his first attempts to mate that he refrains from sexual behavior in the future.

The castrating male is also correct. Dominant toms may fight and chase inexperienced males away from the circle of females, preventing them from mating and psychologically castrating them.

One experiment conducted with monkeys found that dominant males showed an increase in testosterone, the male hormone, within twenty-four hours of interacting with other males, while subordinate males showed a decrease in testosterone levels.

Trauma to testicles or penis may also result in physical sterility. This occurs from fight wounds and from surgery. The perineal urethrostomy is a surgical technique to widen the male urethra at the pelvic inlet to allow the passage of stones and mucus in cats with recurring urinary disease. In this procedure, the penis is amputated and the urethra widened—a sex-change surgery of sorts. Naturally, reconstructing the genital anatomy is a last-resort type of decision.

Sexual Problems (Female)

Pseudopregnancy (False Pregnancy)

Sometimes, the female cat is induced to ovulate by natural breeding or by a simulation of breeding but fails to become pregnant. This occurs when the male is sterile or fails to achieve intromission, or in rare cases when the female is stroked over the back or stimulated when the veterinarian takes a vaginal swab to check the stage of the heat cycle. The body in these cases is tricked into thinking it is pregnant. The cat ceases mating behavior for about forty-five days.

Thankfully, few cats suffering from pseudopregnancy show the breast enlargement and maternal nervous behavior seen in dogs suffering from this condition. Most feline patients require no therapy; for those requiring medication, physical symptoms are relieved with hormones and psychological symptoms with tranquilizers.

Breast Enlargement

A few young females show engorgement of one or several breasts after the first heat cycle. The mammary tissue involved is swollen, hot, and painful. The cause of this condition is high levels of the female hormone estrogen; breast enlargement recurs each heat cycle, making spaying the only logical treatment.

Nymphomania

Female cats exhibiting estrus behavior continually during the breeding season might be called nymphomaniacs. The causation in most cases is cystic ovaries pouring a constant supply of the female hormone estrogen into the cat's system. Nymphomania occurs most commonly in Siamese and Persian cats. Spaying is curative.

Preventing Pregnancy

Feline overpopulation is a major concern. A large percentage of unwanted kittens are abandoned to lives of starvation, disease, and trauma. One study reveals that cats killed along highways are primarily kittens and young adults. At least 75 percent of cats that end up at animal shelters are put to sleep.

An intact female can produce a litter of kittens shortly after she reaches puberty. Approximately eight months later the females in that litter produce kittens, and so on. An individual queen can be the source of over four thousand offspring within seven years. It makes sense to neuter cats, both males and females, before the cycle of unwanted births and before behavioral problems associated with sex and the single cat begin.

Neutering

Spaying (ovariohysterectomy surgery) not only prevents pregnancy, it also significantly decreases the incidence of breast cancer and eliminates the possibility of uterine disease in later life. Intact queens experience mammary tumors at a rate seven times that of spayed female cats, and the majority of feline breast tumors are malignant.

Castration eliminates tumors of the testicles and significantly decreases spraying, roaming, and intermale fighting.

It is best to have the female cat spayed as soon as she reaches physical maturity and before the first heat cycle. Most veterinarians recommend that your female cat be spayed and the male be castrated at six months of age, although some veterinarians and animal shelters are advocating neutering animals as young as two or three months. Your veterinarian should be consulted prior to this time for his or her recommendations for your individual cat.

There is no reason to let a cat have one heat cycle before spaying unless you enjoy staying up all night with an amorous cat. There is no reason to think that your cat should have at least one litter before neutering. It is my liberated view that no cat need bear young to be fulfilled. That goes for women, too. Uh oh, I feel myself climbing on my soapbox. Just one more: There are better ways to teach your children about sex and birth than watching Fluffy doing what comes naturally.

For the surgical procedures of castration and spaying, general anesthesia is required. It is best that your pet be in good health and have received protective vaccinations prior to scheduling surgery. Surgery itself is a stress, both physically and mentally; vaccinations and other medical procedures add to that stress when combined with surgery.

In the female, the abdomen is opened, and both the uterus (uterine horns and body) and ovaries are removed. Your pet cannot get pregnant, nor will she come back in heat following surgery. Several stitches will be required to close the abdominal incision.

It is best to schedule ovariohysterectomy surgery at a time when the cat is not in heat or pregnant, although it is better to spay in heat or within the first three weeks of pregnancy than risk an unwanted litter. After four to five weeks of pregnancy, the surgery poses an increased risk to the mother because of engorgement of blood vessels and loss of fluid during surgery.

In males, a small incision is made in the cat's scrotum, and both testicles are removed. Male cats usually heal well without stitches.

Pets recovering from surgery should be given as much love and attention as possible. Pets with stitches should be confined to the house and closely monitored until the sutures are removed, usually in ten to fourteen days.

Table 24: The Advantages of Neutering

Prevents female heat.

Prevents pregnancy and birth of unwanted kittens.

Prevents diseases of reproductive organs in both males and females.

Significantly decreases incidence of breast cancer in the female.

Decreases objectionable mounting behavior.

Decreases fighting between males.

Decreases incidence of urine and fecal marking.

Decreases roaming, reducing accidents, exposure to infectious diseases, and lost cats.

The Morning-after Shot

An unplanned pregnancy can be prevented with a morning-after, or mismate, shot. An injection of the female hormone estrogen within forty hours of mating will usually terminate the pregnancy. The hormone blocks fertilized eggs from entering the uterus. Because of the potential for harmful side effects from the drug, such as the increased risk of uterine infections, most veterinarians are reluctant to recommend this means of terminating pregnancy, especially on a routine basis.

Abortion

Newer drugs in the prostaglandin category cause abortions in the cat. The drugs must be given after day forty of pregnancy; they work by causing a cessation of progesterone (the pregnancy hormone) production by the ovaries. Because these drugs have not been approved for clinical use in cats, most veterinarians are reluctant to use or recommend them for feline abortions.

Birth-control Pills

There are two birth-control pills on the market for dogs. A progestin, megestrol acetate, is the same drug cited in earlier chapters for treating various feline skin and behavioral problems. This drug can be given to cats to prevent them from coming into heat. The drug has not been officially approved for cats and offers potential side effects as serious as diabetes. The other birth-control pill for dogs, containing the drug mibolerone, should not be given to cats because of its toxicity in this species. As you can see, birth-control pills for cats are not a viable option as yet.

The Future

What is needed is a quick, inexpensive, long-lasting, and safe means of preventing cats from reproducing. Both intrauterine devices and chemical means of sterilizing pets are being investigated. Synthetic hormones and contraceptive vaccines are hopefully on the horizon. In the meantime, the quality of life for our cats depends upon early neutering for all animals not intended for breeding.

Questions

Dear Dr. Whiteley,

My cat Bouncer has been diagnosed as having a skin condition called stud tail. The big brute was neutered at six months old. How could he have stud tail?

Willie in Chicago

Dear Willie,

The condition of stud tail was poorly named, for it can occur in neutered males and females. The oily skin glands located on top of the tail become overactive and "blackheads" form in the pores. Since it occurs most often in intact males, neutering is recommended. Otherwise, frequent shampooing of the area with a benzoyl peroxide shampoo, good nutrition, and daily grooming control what is usually a chronic condition.

Good luck!

H.E.W.

Dear Dr. Whiteley,

My darling Valentino, a registered Persian, has started a most disgusting habit. He often accompanies me to bed, where he used to lie down and sleep like a nice polite fellow. Lately, however, he has begun this grasping motion with his paws on the soft bedspread and masturbating against the pillow or whatever else I happen to be leaning against at the time. What could make a nice cat do such a thing?

Disgusted Lady in Memphis

Dear Disgusted,

I assume that you and Valentino have had a close relationship since he was a kitten. Often a kitten that bonds closely with his human mother carries those feelings of attachment for her into the sexual area when he gets older.

You did not mention if Valentino was neutered. If not, I suggest that it is time to have the surgery done. If he is neutered already, you must take steps to put some distance in the relationship. You may have to make your bedroom off-limits for a while. Offer Valentino one of those cozy heated beds and a saucer of milk in an area of the house located away from your room. If he meows at the door, ignore him. Reward good behavior—sleeping quietly in his own bed.

Best wishes!

H.E.W.

12

Feline Parenting

Dating Services

Ah, spring—the time of the birds and the bees, the flowers and the trees, and the personal ads—lovely Persian female with silver hair and blue eyes desires dark and handsome Persian male with flat face, winning personality, and championship background. Object: romance and parenthood.

Romance and parenthood, however, should be more than a spur-of-the-moment affair. An animal caretaker must be convinced that he or she is improving the breed before setting up a feline "date." Litters from unpedigreed parents cannot be registered, and homes for the offspring may be difficult to find. Congenital defects such as retained testicles and polydactylism (excessive number of toes) can be passed to offspring that few are willing to adopt. Research the background of the romantic pair thoroughly.

Know your own pet's strong and weak points before selecting a mate. Breed a cat with a certain weakness to a mate with a corresponding strength. Look at littermates and parents of the prospective couple to get an idea of physical and behavioral family traits. Consult a pedigree service to inquire about the colors and championships of the cats' ancestors. The Cat Fanciers' Data Center (77 Essex Avenue, Montclair, NJ 07042) provides extensive information about the background of registered cats of every major breed.

Avoid inbreeding. Animal breeders classify the mating of father and daughter, mother and son, or full brothers and sisters as inbreeding; genetically, an individual is as closely related to a sibling as to an offspring. The mating of close but less direct relatives, such as cousins, is called line breeding. Inbred- and line-bred pairings intensify faults as well as attributes. The probability that the mating between two individuals will result in offspring with congenital defects is directly correlated with the degree of inbreeding.

Beware of genetic disorders prevalent in breeds that have been highly inbred. Cats and their relatives suffering from genetic disorders should not be mated. Genetic counseling advertised in many cat magazines might be worthwhile if you know little about your cat's breed. The Cat Fanciers' Association (1805 Atlantic Avenue, Manasquan, NJ 08736), the largest of several organizations devoted to registering cats, can provide breed information.

A cat that appears normal may carry a gene for a genetic defect.

If this animal mates with another normal-appearing cat carrying the defective gene, there is a one in four chance that a kitten from the pair will receive a defective gene from both parents, thus manifesting the genetic problem. When a single gene mutation occurs, it may take only one parent carrying the defective gene to produce signs of the genetic disorder. There will then be a one in two chance for an offspring to suffer the defect.

Table 25: Breed Predisposition to Genetic and Congenital Disorders

Chediak-Higashi syndrome (disorder of eyes and skin) occurs in Persians with green or yellow eyes and "blue smoke" hair.

Hereditary baldness: Siamese, cats bred as sphinx or Canadian hairless cats.

Spina bifida (failure of vertebrae of back to close normally around the spinal cord): Manx.

Deafness, decreased fertility, vision, immunity: blue-eyed white cats, except Burmese and Siamese cats with this coloration.

Hydrocephalus (excess fluid within brain): Siamese.

Cleft palate/cleft lip: Siamese.

Crossed eyes: Siamese.

Turning in of lower eyelids, causing irritation to eye: Persian.

Deficient tear production, predisposing cat to eye infections: Burmese, Abyssinian.

Abnormal shortening of face with absence of nasal tissue (lethal condition): Burmese.

Corneal dystrophy (degeneration of cornea of eye): stump-tailed Manx.

Psychogenic alopecia (self-induced baldness due to licking and hair pulling): Siamese, Burmese, Abyssinian.

Pyloric dysfunction/megaesophagus with accompanying regurgitation and vomiting: Siamese.

Absence of anus (external opening of rectum): stump- and rump-tailed Manx.

Dilated heart: Siamese and Burmese.

Porphyria (brownish discoloration of teeth, bones, urine): domestic shorthair, Siamese.

Familial amyloidosis (kidney disease): Abyssinian.

Metabolic-enzyme deficiency (large head, small ears, abnormalities in gait): Korat.

Breeding Management

Before breeding, both males and females should be examined by a veterinarian for physical problems that might affect pregnancy, plus checked for parasites and inoculated against infectious diseases. Both deworming and vaccinations should be complete prior to breeding. Parasites can be passed to kittens during pregnancy, and many modified live-virus vaccines have the potential to cause birth defects in kittens if given to the pregnant queen. Because feline leukemia virus can be passed from queen to kittens during pregnancy, cats that have not been tested and vaccinated against this disease should be tested prior to breeding.

Most females ten to twelve months of age are old enough for breeding for the first time. Queens older than ten years should be retired from breeding. Females that reach puberty later than eighteen months, those that fail to conceive after three breeding cycles, and those that repeatedly deliver weak or dead kittens should also be retired from future breeding.

Queens in heat (estrus) should be taken to the stud cat's home or cage for breeding and allowed to mate two to five times daily for as long as they will accept the tom. Queens should not raise more than two litters yearly for optimum health of both mother and kittens.

Pregnancy

Multiple matings trigger an increase in reproductive hormones, causing the release of eggs from the ovaries about twenty-four hours later. These eggs, which average four in number, must be fertilized within twenty-four hours of release from the ovaries for pregnancy to result.

After mating and ovulation, millions of sperm implanted by the male during copulation travel up the female reproductive tract to the oviducts (Fallopian tubes), where the eggs wait. One sperm cell penetrates the covering of one egg to form the beginning kitten. The genetic material of the individual is sealed because the egg is altered during fertilization to prevent penetration by additional sperm. If the queen mated with several toms, sperm from another father can penetrate a different egg, forming an offspring that is a half-sibling to the first.

Usually, the queen ceases to show signs of heat twenty-four hours after ovulation, and heat behavior does not resume until three weeks after kittens have been weaned. Occasionally, however, a pregnant queen will show signs of heat during the third to sixth week of pregnancy, presumably a result of secretion of the hormone estrogen from the placenta. If the queen is bred several times during this heat cycle, a second litter of kittens differing in age from the first can develop in the uterus at the same time. This rare situation has resulted in a queen delivering full-term kittens several weeks apart.

Although not a common occurrence, an increase in estrogen levels near or soon after birth can also result in the queen showing signs of heat soon after the delivery of kittens. This queen exposed to mating toms suffers the increased risk and stress of constant pregnancy except during that quiescent period in winter.

The average length of pregnancy (gestation) in the cat is sixty-three to sixty-five days. Larger breeds tend to have longer gestations, often approaching seventy days. The date of breedings should be recorded carefully so that pregnancy can be monitored and birth plans made. During the nine weeks of gestation, developing kittens change from a single-cell egg measuring approximately one–two hundred fiftieth of an inch to a four- to five-inch newborn.

Six days after fertilization, fertilized eggs, which are undergoing cell division and traveling down the oviducts, arrive at the uterus. Implantation of embryos into the walls of the uterine horns occurs at days eleven to twelve of pregnancy. Improper development of the growing embryo during these first two weeks is usually deadly, and most abortions occur during this critical time.

The next phase in the development of unborn kittens occurs from days twelve to twenty-four, the time that all organ systems are established within the embryo. From day twenty-four to term, the fetus undergoes rapid growth and fine-tuning of all organ systems. At about day twenty-five of gestation, half-inch fetal kittens begin movement. Kittens born before eight weeks of gestation are premature and rarely survive.

Detection of Pregnancy

Signs of early pregnancy may include a decrease in appetite and activity, with occasional mild vomiting; however, most cats do not display these symptoms of morning sickness so familiar to human mothers. Two and a half to three weeks after mating, the nipples may become darker pink and more erect than before. Of course, later in pregnancy the abdomen enlarges.

Currently, there are no urine or blood tests available to detect feline pregnancy. The most common means of feline pregnancy diagnosis is palpation by your cat's veterinarian. By two and a half to three weeks, the growing fetuses can usually be felt by an experienced clinician. Pregnancy can also be determined by radiographs (X rays) after thirty-eight days of gestation. Earlier exposure to X rays may be damaging to the developing kittens; after five weeks of gestation, the fetal skeletons are detectable on radiographs and the procedure is less risky to the kittens. Ultrasound, now more commonly available at veterinary hospitals, is a reliable means of detecting pregnancy, usually by day twenty-four after matings, and of monitoring the progress of fetal development. The fetal heartbeat is detectable on ultrasound by day twenty-four.

Care of Pregnant Queen

If problems occur with other cats in the household, it may be advisable to isolate the prospective mother from the troublemakers. Most cattery owners provide a separate area or cage for pregnant queens.

Diet is an important consideration during pregnancy. The total caloric requirements for the pregnant mom increase 25 to 50 percent during pregnancy (diet should supply 1700 digestible KCals/lb); protein requirements are also higher (30 percent highly digestible protein). During milk production (lactation) the caloric requirements increase to twice the female's normal needs. Increasing the regular diet may suffice during early pregnancy. I recommend a growth/lactating diet during lactation and the last half of pregnancy. The queen should be fed free choice or frequent smaller portion meals. A queen receiving a nutritionally complete diet designed for this life stage of pregnancy and lactation does not need supplementation such as vitamin/mineral tablets or table foods.

After queening, the female does not return immediately to prepregnancy weight; excess fat is helpful, especially if mom has a large litter to feed, for providing energy necessary for milk production. Mothers tend to lose weight when lactating. Milk production is linked directly to maternal body weight; heavier queens tend to give more milk.

Weaning of kittens starts when they approach four weeks of age. Hopefully, they will follow their mother's lead by consuming her growth/lactating diet, which supplies the high-caloric, high-protein needs of growing kittens. The food may be moistened to make it easier for kittens to eat.

Toward the end of pregnancy, a warm, soft, and secluded nesting box may be provided for the delivery and housing of kittens following birth. The sides should be high enough that kittens won't roll out but low enough that the queen can come and go over the sides easily. Nesting material should be soft and disposable, and the box should be located in a warm area of the house, away from traffic and noise. Of course, the best laid plans go awry when the queen ignores the nesting box you've gone to such pains to provide, opting for the darkest and most cluttered area of the closet or the cobweb-infested underside of the bed.

Long-haired cats may be prepared for the birthing process by clipping long hair surrounding the vulva and nipples.

Labor and Birth

The act of giving birth in cats (queening) can be divided into three distinct stages. Stage one heralds the coming birth and lasts from a few hours to one day prior to the onset of obvious contractions. The breasts fill with milk, which can be expressed from each gland. There is an increased restlessness, pacing, vomiting, and meowing at this time. The queen may scratch and fuss with the bedding in her nesting box. She may squat in her litter box without voiding in response to early contractions. In preparation for labor, the queen will lick and groom the nipples and vulva. She may lose her appetite or may eat voraciously until and during the birth of kittens. Some queens will show a drop in rectal temperature from the normal of 101.5 degrees Fahrenheit to 99 to 100 degrees.

The dependent queen will seek out her human caretaker and solicit attention at this time. She seems reluctant to assume the upcoming role of mother alone and is comforted by nearby humans. There is anecdotal evidence that the human-dependent cat tends to have prolonged labor.

Other cats behave in an opposite manner, seeking privacy and seclusion, particularly from humans and other animals. Within reason, give your cat what she wants. Hopefully, you can reassure the dependent kitty without sleeping in her nesting box with her, and you can quietly monitor the progress of the shy one without disturbing her sense of privacy.

Stage two is the birth of kittens; the average litter is three to five kittens. In most queens, labor lasts from three to six hours; although occasionally a mother may deliver a kitten or two and cease all signs of labor. She nurses and eats as if she is finished with the whole process; the delivery of a second set of kittens proceeds in twelve to twenty-four hours. This interruption of labor must be differentiated from birthing problems, usually expressed by continued contractions and discomfort.

Contractions can be seen as well as felt, as the queen's uterine horns work to expel kittens. She may pant and groan or cry, especially as the first little fellow is being born. She may lie on her side or on her chest, or get up and assume a squatting position.

Prolonged labor with contractions signals a problem in cats. After hard contractions begin, the first kitten should arrive within a few minutes to, at the most, an hour and a half. The same goes for successive kittens.

If the water bag of the first kitten breaks, the queen will put her head under a rear leg and lick vigorously at her rear. If the water bag is intact, the shiny balloonlike sac will be noticed at the vulva before the kitten is expelled. If the bag was torn as the kitten entered the birth canal, the head, legs, or tail will be seen first. Breech and headfirst positions are considered normal presentations.

As the kitten is delivered the queen severs the umbilical cord attaching the kitten to the placenta with her teeth and licks the newborn; this removes the kitten from its sac, if it is still intact, and stimulates respiration. Some mothers, however, stop the process of taking care of the infant to clean themselves and eat the placenta. If the queen abandons the kitten before the sac has been removed, cord severed, or breathing stimulated, you must step in to assist the newborn.

Wash your hands before handling the newborn. Remove the sac from around the kitten, exposing his nose to air first. The cord can be tied with string approximately one inch from the kitten's body and severed below the knot. Rub the kitten with a clean diaper or towel. If the infant is not breathing, remove fluid from the nose and mouth with a bulb syringe. Continue rubbing to stimulate breathing.

The third stage of labor is delivery of placentas. This brownish tissue is expelled with each kitten or shortly thereafter. It is not necessary that the queen eat this tissue, and if placentas can be removed without disturbing the new mother, the diarrhea that often accompanies eating of the placentas may be prevented. In the wild, eating the placenta serves a worthwhile function: It provides nutrients, allowing the mother to spend more time with the young than if she had to hunt for food immediately, and also serves to keep the nest clean. In the midst of birth, it is difficult, if not impossible, to keep up with the placentas. However, an accounting of each placenta is helpful for the veterinarian who is trying to diagnose post-delivery problems. A placenta remaining in the uterus predisposes the queen to infection.

Birthing occurs frequently at night or on the weekend, making the decision to consult your pet's veterinarian a judgment call. I remember being called late at night by a neighbor who wanted to know if she should call and disturb her regular vet with her questions about her cat's new kittens. I don't recommend this roundabout way of deciding when to call the veterinarian—not if you want to remain friends with your neighboring vet.

Of course, the best time for the first consultation is before breeding and potential problems crop up, as discussed in the section on breeding management (page 214). Cats with the potential for problems should be seen before the onset of labor. For cats without prior problems, the veterinarian should be consulted as soon as serious problems are detected (see Table 26).

For cats with a normal delivery, the mother and kittens should be checked the next day or Monday, if queening occurs during the weekend. The kittens and mother can travel to the veterinary hospital in a small covered box. If your veterinarian makes house calls, this is a time to schedule a home visit.

Table 26: When to Consult the Vet

Cats having a history of pelvic fractures, gross obesity, chronic illnesses, or previous cesarean sections: Consult before onset of labor or, better yet, before breeding.

Passage of bright red or greenish-colored discharge from vulva without onset of labor.

Prolonged gestation (over sixty-eight days).

Prolonged contractions without producing a kitten (over ninety minutes).

Kitten stuck in birth canal.

Weak, diminished contractions without producing kitten.

Mother or kittens depressed, cold, with pale mucous membranes (can be determined by examining the color of gums, tongue, membranes surrounding eyes) during or after birth.

Uterine prolapse following birth: presence of one or two turgid tubular structures extending from the vulva.

Mother not eating twenty-four hours following queening; mother with a body temperature over 103 degrees Fahrenheit; mother showing malodorous discharge from the vulva.

Mother having no milk or refusing to let kittens nurse following birth. Kittens too weak to nurse following birth.

Female Parenting

Nursing Behavior

Newborn kittens attempt to nurse within one to two hours of birth. The queen may allow the kittens to nurse between births or may refrain from nursing until the last baby is delivered.

The first milk (colostrum) contains antibodies that protect newborn kittens against infectious diseases. Adequate nursing during the first day of life ensures that kittens receive this antibody protection, and also helps to keep kittens warm and to maintain their blood pressure.

The queen stays with the litter for the first twelve hours after birth, spending most of this time nursing infants. She is providing approximately one milliliter of milk to each infant per hour.

The queen lies on her side with her legs and body encircling the kittens. Mother licks the kittens, orienting them toward her breasts. Queens have eight nipples, but usually only six are functional; six is an adequate number for an average litter of four to five kittens.

The immature, functionally blind and deaf kittens are also attracted to nipples by smell. Within three days, most kittens (about 80 percent) have selected a specific nipple that they consider theirs for approximately a month of nursing; rear nipples are preferred. Attraction to a specific nipple serves several purposes: It lessens claw injuries among littermates and provides maximum mammary stimulation, hence more rapid completion of nursing. If a specific teat is not sucked for three days, milk production in that mammary gland ceases. Kittens perform the "milk tread," rhythmic alternate movements of forepaws against the mother's breast. The milk tread is believed to stimulate milk flow and push the mother's skin away from the kitten's nose. Purring by both queen and kittens often accompanies nursing.

The queen spends approximately 90 percent of her time with kittens the first week following birth, as infants are totally dependent upon mother at this time. Seventy percent of this togetherness time is spent nursing. By the second week of life, kittens are consuming five to seven milliliters of milk (one teaspoon equals five milliliters) per feeding and have doubled their birth weight. An average kitten weighing three and a half to four ounces at birth should be a heavyweight of one pound by the third week.

There is a slight difference in birth weights between male and female kittens, with females being slightly heavier at birth and growing slightly faster during the first week of life. By six weeks or weaning time, males have caught up in weight and will soon surpass female littermates in size and poundage.

Grooming and cleaning take up most of the mother's time not spent nursing during those early weeks following queening. Kitten elimination and cleanliness of the nest is maintained by the mother licking the genitals and ingesting feces and urine from newborns. Mother also retrieves part of the water she invests in milk production in this way. After three weeks, the kittens are mobile enough to defecate and urinate away from the nest, and follow their mother's lead in doing so.

For the first three weeks, nursing is initiated by the queen. After the second week, the kitten's eyes and ears are functional, and they are more independent. By the third week, kittens are walking and even running, and initiate most nursing sessions. The mother generally cooperates by lying down and exposing her nipples.

By the fifth week, kittens are climbing and might be considered equivalent to a toddler in the terrible twos. Mother is spending only 16 percent of her time with kittens. When she sees kittens approaching for a bite to eat, she runs and hides, and who can blame her? This behavior might be termed the empty nest syndrome.

In the wild, cats do not wean their young, but the milk has reduced nutritive value after twelve weeks. By this time, kittens should have been taught by mother to hunt and fend for themselves.

In one experiment, kittens raised by an artificial mom (made with a soft surface to snuggle against and nipples from which to suckle) were compared with kittens raised by a natural cat mother.

The artificial mom did not interact with kittens in the same way as the natural mother, and kittens remained socially naive and immature. The artificial mother made no attempt to reject kittens, and they, therefore, remained dependent and made no attempt to wean themselves. Distancing of offspring serves their best interests as well as the mother's.

Retrieving Behavior

Along with providing nutrients through milk and cleanup duties by licking babies after nursing, good cat mothers must transfer their own thermal energy to kittens until they can regulate their own body heat, beginning at about ten days of age. It takes an increase in body fat for the kittens to maintain their body temperature; by four weeks, kittens provide their own thermal energy.

Feline mothers must stay with young kittens a large percent of the time, and must make sure that infants stay close and warm. If an infant falls out of the nest, and some cling to a nipple and get dropped elsewhere, the queen must respond to the kitten's plaintive cry and retrieve the lost one before he becomes chilled and dies.

The queen retrieves the kitten by grasping the little fellow by the nape with her teeth. The youngster quiets down and remains immobile while being carried back home to safety. This behavior is reflexive, and an adult cat can sometimes be carried by grasping him by the nape.

If there are disturbances or perceived danger in the immediate environment of the nest, the astute mother packs up the brood and leaves. This tendency to move is greatest three to five weeks after queening.

If a kitten falls out of the nest and his mother fails to retrieve him, you must warm him and present him to his mother. The kitten can be warmed by placing him on a carefully monitored heating pad or in a small box next to a bottle filled with warm water or with an overhead light bulb. Kittens raised away from the nest should have an ambient temperature of 75 to 80 degrees Fahrenheit. If the mother fails to accept and care for the baby after he is placed back in the nest, he must be raised as an orphan.

"Mothering"

Interestingly, the cooperative queen may accept young from another cat or even from another species to her breast during the first week after queening. If a strange kitten is presented rear first to the prospective mother, the queen often responds automatically by licking the kitten's genitals and anus; if the kitten is presented headfirst, the queen may act aggressively toward him. The queen is most likely to foster young not her own if they are introduced to her at younger than one week of age or at weaning age, when mother might be more inclined to share milk or prey.

When two or more queens give birth at approximately the same time and are housed in close proximity, they may take turns nursing each other's kittens. An aggressive female, whether she has kittens of her own or not, may try to take the kittens away from the more timid mother.

Prior association between a mother and her offspring controls the mother's recognition of her young. Offspring removed at birth and reared apart from their biological mother are not usually recognized by her. Recognition of one's own young serves to enhance the bonding between mother and offspring, which is added insurance that helpless kittens will survive.

The bond is formed between kittens and queen primarily by nursing and licking during the first week after queening. Of course, nursing and licking stimulate all of the senses, so the mother feels, sees, hears, smells, and actually tastes her kittens as she ingests their urine and feces.

In humans, preliminary studies indicate that during the first three days following the birth of their infant, new mothers show an increase in the hormones associated with mothering (estrogen, prolactin, pregnancy steroids) and an increased ability to detect infant-related odors, sounds, and the feel of their own young.

In one experiment, researchers blindfolded human mothers and had them stroke the backs of three newborns' hands in an effort to determine if the new mothers could identify their own infant from touch alone. Seventy percent of the mothers correctly chose their own infant—nearly twice the 33 percent expected from random guessing. Many of the mothers said that they used the texture or temperature of the infants' hands to make their determination.

In a separate study, human mothers were able to select their own infant's smell (from newborn T-shirts) from four other same-age infants at greater than chance odds. As few as two hours spent with the baby resulted in successful recognition. Mothers are also very adept at picking out the cry of their own baby from those of others in the hospital nursery.

In another experiment using rats, it was found that microscopic changes occur in the brains of maternally behaving rats that are different from those of nonpregnant or nonlactating rats. These tiny brain differences occur in conjunction with the hormones of pregnancy and with the sensory stimulation induced by the presence of rat babies. The changes are reversible and after weaning are no longer detectable.

There is a basic capacity to respond maternally to one's young, but there is a learned component, too. Rhesus monkey mothers that were themselves deprived of maternal stimulation during their early developmental years often failed to care for their young at the end of their first pregnancy. However, these same mothers overcame their earlier deprivation and displayed relatively normal maternal behavior toward their young after the second and third pregnancies.

It seems that nature has designed mothers to be mothers, and when mothers fail to perform this role properly owing to their own inexperience or deprived backgrounds, teaching, understanding, and experience can go a long way toward developing maternal behavior.

Maternal Aggression

A queen often shows aggressive behavior toward toms before queening and while kittens are nursing and dependent upon her protection. This is because male cats have been known to kill newborn kittens. The protective mother will also become aggressive toward humans, cats, and other animals that she perceives as a threat to her litter.

Studies in rats show that maternal aggression evidenced by threatening behavior toward others is first observed during mid-pregnancy and increases as the end of pregnancy nears. This aggressive behavior peaks during early nursing of the young. By the time the offspring reach weaning age, the aggressive behavior wanes. This aggressive behavior is believed to be hormonally driven.

MATERNAL CANNIBALISM—Eating of kittens by the mother occasionally occurs when she gives birth to a larger than normal number of young and when weak, ill, or deformed kittens are present in the litter.

In the wild, the behavior makes sense under the survival of the fittest principle. When food is scarce and there are too many mouths to feed, reducing the size of the litter increases the likelihood that those remaining will thrive. Using her own kittens as a source of food lessens the necessity that the mother will have to leave the nest to hunt. Eliminating ill kittens reduces the chances of spreading disease to the others. Eating deformed kittens before they die increases the resources available to the others and prevents the nest from becoming soiled with a dead body that might attract scavengers.

Cannibalism is also a reality in homes where resources are readily supplied by the owner and accounts for 12.5 percent of the preweaning deaths of kittens. Nervous individuals seem more prone to this type of behavior than laid-back cats. Stress often triggers what the queen perceives as a threat to the nest, and as we've discussed in earlier chapters, stress can include noise, overcrowding, malnutrition, visitors, and other types of disruption within the queen's environment. The removal of sources of stress, good nutrition, and tranquilizers may be helpful for the distressed mother.

Male Parenting

Parental care evolves if the reproductive benefits outweigh the costs. For toms, reproductive success is measured in terms of the number of females impregnated, not in the number of kittens nurtured.

Although exceptions occur, most toms do not stay with or parent a litter and hence fail to recognize kittens as their own. If the male lives in close proximity to the female and kittens, he may develop an interest in and preference for familiar kittens.

In the wild, male cats may commit cannibalism when they encounter a queen with kittens. By killing the young, the female will come into heat sooner, and the aggressive male can now have a chance to sire another litter by her.

A new acquaintance recently told me of an early trauma that soured her on cats. When the woman, now in her sixties, was six years old, the barn cat had a litter of kittens on a cold and snowy March day. The girl's father noted that the queen had moved the kittens under a bushel basket he had left outside the barn. The girl ran out one morning to peek at the kittens and found all of them with their heads bitten off. The father cited an old wives' tale about toms killing March kittens. The woman remembers the incident vividly and still harbors a grudge against a species that would behave in such a manner toward its young.

I had never heard the tale about the March kittens, but in a way it makes sense. A March litter would be the first of the season, and a traveling tom bent on spreading his genes around to as many females as possible might choose infanticide as a way of making the mother available to him.

I suppose one can say that this is nature's way of ensuring the survival of the most aggressive and dominant cats. Still, I recommend that queens with kittens be kept safely away from strange tomcats until the kittens are weaned. This is also another good argument for neutering wandering toms.

Reproductive Failures

It seems to me a miracle that in most cases, young are conceived, carried to term, delivered, nursed, weaned, and raised to maturity to start the cycle all over again. Many things can go wrong at any stage of the cycle that would prevent a kitten from reaching adulthood; this is one of the reasons that nature designed the cat to have multiple young and multiple pregnancies during a season, allowing only for reproductive rest during winter when the chances of survival reach even more dismal odds.

Failure to conceive or deliver viable young can have many causes. Trauma, nutritional imbalances before or during pregnancy (taurine deficiency, for example), physical abnormalities of the genital tract, physical or psychological stress, exposure to toxins, hormonal imbalances, malformed fetuses, and organ dysfunction are just a few of the reasons that reproduction fails.

Infectious agents also work against healthy reproduction. Toxoplasmosis, a parasite, and feline panleukopenia and viral rhinotracheitis, two diseases for which we normally vaccinate cats, can cause abortion and death of kittens. Feline leukemia virus and feline infectious peritonitis virus affect multiple organ systems, of which the reproductive system is one. See Chapter 10 for a detailed discussion of these diseases. To prevent reproductive failures from disease and other agents, keep catteries clean, space adequate, nutrition at an optimum plane, and adhere to good breeding management, including vaccination programs and parasite control.

Question

Dear Mr. Whiteley,

My mother cat had four kittens last year. When we gave the babies away at six weeks of age, Mama, as we call her, was filled with milk and terribly miserable. She is expecting again, and I am going to have her spayed as soon as possible after the kittens are weaned. What, if anything, can I do so she won't have that much milk?

Ruby in Phoenix

Dear Ruby,

Mama sounds like a good cat mother. Perhaps she needs help in enticing her kitties to eat on their own. Starting at three weeks of age, offer the babies a saucer filled with KMR (Kitten Milk Replacer), marketed by Pet-Ag. This is the same formula used to feed orphan kittens, and is available at veterinary hospitals and pet stores in canned or powdered form. In the next week or two, offer a kitten chow moistened with Kitten Weaning Formula (Pet-Ag). Hopefully, you'll have those kitties nursing very little by the time they go to new homes.

If Mama is engorged with milk after the babies leave home, limit her access to liquids and food for twenty-four hours. The pain of breast swelling is eased by rubbing with cocoa butter and perhaps binding snugly with an Ace bandage. Be careful that you don't wrap the bandage too tightly, and monitor it closely. If this doesn't work, your veterinarian can give her a couple of injections that will help.

Last, but certainly not least, make an appointment for Mama's surgery as soon as she is dried up and before she has a chance to become pregnant again.

My best!

H.E.W.

13

Aging

According to the experts, our bodies age because of changes in cellular genetic material, leading to the death and impaired function of body cells. These changes occur spontaneously and in response to radiation and chemicals. If we can find ways to delay, prevent, or reverse these changes, aging could be delayed, reversed, or stopped completely.

If we found a way of preventing aging and death, however, where would we all go to find a quiet spot for a picnic? A solitary walk in nature? How long would the line be at Furr's Cafeteria after church on Sunday? Would everyone, regardless of age, want to take up motorcycle racing, sky diving, and bungee jumping?

Finding the fountain of youth that prevents or reverses aging has fascinated humans throughout the eons. Cats, however, never worry about aging or death and spend their lifetime in the present —eating, grooming, sleeping, playing, hunting, raising young, or mating. Still, if a fountain of youth is ever found, it will benefit cats as well as people.

I'm convinced that the rewards of "fountain of youth" or aging research will not be interminable physical life but the prolongation of mental and physical well-being so that we, cats included, can live our limited lives to the fullest.

Longevity Factors

The average life span for a species tends to increase with the size of the species. For example, elephants—the oldest of which is fifty-seven years—live longer than cats, which live longer than mice. The exception is man, who may live to a maximum of 110 years.

Within a species, however, the smaller sex tends to live longer than the larger sex. In humans and cats, the female generally outlives the male. In hamsters, the female is the larger gender, and the male is longer lived.

Table 27:
Maximum Life Spans (in years)

Man: 110	Cat: 36	Horse: 50
Elephant: 57	Mouse: 4	

The longevity of individuals within a species is genetically determined. Long-living parents tend to produce offspring that live to a ripe old age. Hybrid vigor is also a factor. Genetic diversity of offspring produced by mating two different but pure lines (breeds or varieties) increases strength and resistance to disease in the resulting organism, and it doesn't matter if the organism is a carrot or a cat. Breeding of closely related individuals tends to produce shorter-lived offspring.

Other Factors

Nutrition plays an important part in health and longevity. Extremely malnourished animals suffer slower wound healing, impaired disease resistance, increased susceptibility to cancer, and thus, early death.

Overfeeding to the point of obesity also has a damaging effect on health and longevity. Most damage is due to a reduction of cellular immunity, organic dysfunction, and increased susceptibility to autoimmune diseases.

Moderate caloric restriction, however, has a positive effect upon extending life and retarding disease development. Dr. Roderick Bronson, a scientist at the United States Department of Agriculture Human Research Center on Aging, found that laboratory rodents fed 40 percent fewer calories than normal-fed controls lived 29 percent longer and suffered less from cancer and other diseases. This correlates well with reports that the longest-lived humans reside in mountainous regions where diets are low in calories and life is hard and strenuous. People in these remote regions are also free from crowding, which produces a rapid exchange of disease-causing viruses.

I am beginning to think that many felines and humans, myself included, should change our mindset from "living to eat" to "eating to live" and just barely, at that. Moving to Outer Mongolia and carrying water and wood might also help.

Lack of exercise contributes to early aging and death. Adequate physical activity is important to maintain muscle tone and lean body mass, to enhance circulation, and to improve waste elimination.

Preventative medicine—dental care, parasite control, vaccination against infectious diseases—and neutering all contribute to our cats' longevity.

How Old Is Old?

As I grow older the age I consider "old" increases. It's not just me; old is actually getting older for all of us, cats and people, because of progress in medicine and health care and "fountain of youth" research. Children born today can expect to live at least seventy-five years; the average life expectancy in 1900 was forty-seven years, and since that's my age now, I'd be making out my will instead of writing this chapter if I'd been born at the turn of the century.

Only in recent years has domestication made growing old a possibility for our feline companions. In the not too distant past, cats starved or froze to death or fell victim to disease or trauma before they reached the age when dry skin and failed hearing and sight became realities.

The life expectancy of a well-cared-for indoor cat is fourteen years; the recordholder for feline longevity lived thirty-six years. Our cats are now living long enough to show the bodily changes associated with aging.

Age, like most things, is relative. Some people maintain energy, both physical and mental, and a sense of wonder and adventure that continues throughout life, while others become rigid and passive at an early age. "Young" better describes a ninety-year-old who is excited about life than a thirty-year-old who has lost interest.

It's the same with cats—one cat is playful and youthful at twelve, and another looks and acts elderly at seven. Individuals do not age at the same rate or in the same manner. Table 29 lists many of the body changes associated with aging. An individual may show some of these changes and seem miraculously resistant to others.

The consequences of aging may create opposite effects. For instance, reduced activity and metabolism may contribute to obesity in one cat, while decaying teeth, dry mouth, and reduction in the senses of smell and taste may result in loss of weight and anorexia in another.

The two oldest cats that I have known were twenty-one and twenty-two years old (equivalent to 100 and 104 human years) when they both developed what became fatal kidney failure. These elderly cats were purebred Siamese and apparently had failed to read that inbred breeds, like Siamese, are not as long-lived as the alley-cat variety of feline. Showing the muscle-wasting characteristic of elderly cats, both were thin, bony, and sagging, and neither had teeth. Of course, each of these cats had received the ultimate in love and care from their human caretakers.

When comparing the physiological age of cats with that of humans, you must take into account that kittens reach puberty between six and eight months of age, whereas boys and girls reach puberty between twelve and thirteen years. The formula that multiplies each cat year by seven is inaccurate, especially when comparing young and old cats to humans. When using the figures of age comparison given in Table 28, ages at puberty and at death of cats and humans correlate more closely.

Table 28:
Feline/Human Physiological Age Comparison

Cat	Human	Cat	Human
6 months	12 years	8 years	48 years
1 year	15 years	9 years	52 years
2 years	24 years	10 years	56 years
3 years	28 years	11 years	60 years
4 years	32 years	12 years	64 years
5 years	36 years	13 years	68 years
6 years	40 years	14 years	72 years
7 years	44 years		

Care of the Aged Cat

Feeding

Metabolism slows at the same time that most pets become more sedentary. If the energy supplied by the diet of cats is not reduced, the cat may become overweight. To counteract a pet's tendency toward obesity, reduce the amount of food or feed a prescription reducing diet. Most veterinary nutritionists recommend reducing the amount of food by 20 percent at age seven, if your pet needs to battle the bulge. Increase play and exercise periods. See Chapters 6 and 7 for details.

Some elderly cats have the opposite problem: They stop eating because of diseased teeth and gums or because their senses of smell and taste are impaired. These cats may lose weight and condition. Feed the reluctant eater several small meals of palatable and easily digestible food; heating the food helps to release the odor, which may entice kitty to eat. If decreased saliva production causes the cat to suffer from a dry mouth, add water to a dry food or feed a canned diet. See Chapter 9.

Aging of internal organs accompanies aging in general, and often older cats suffer from kidney, liver, and cardiac insufficiency. By middle age, seven or eight years for most cats, routine laboratory tests and physical examinations are indicated to monitor organ function. Organ insufficiency detected early can be treated with medication and *diet.*

The hair coat of older cats often becomes sparse and dull. Regular brushing and a high-quality diet with increased levels of vitamins A, B_1, B_6, B_{12}, E, the mineral zinc, and unsaturated fatty acids may be helpful to maintaining the health of hair and skin. Consult your cat's veterinarian about nutrition for the older cat.

Grooming

Grooming, including hair, teeth, and nail care, is especially important for the older cat. The older cat should be brushed often, daily in the case of long-haired cats. If fecal matter soils hair around the tail, trim hair with scissors and wash the area gently with soap and water. The nails should be trimmed every couple of weeks to prevent splitting and breaking. Dry and hard footpads can be massaged with cocoa butter. Teeth care should include cleaning at the veterinary hospital and brushing at home. A soft diet is indicated for cats that have lost most of their teeth. See Chapter 4 for details.

Hormones

Hormones are not an elixir of youth for any of us, but may be indicated for certain older individuals. Spayed females suffering from bladder incontinence are often helped by the administration of estrogen. Overly lethargic cats may be stimulated to activity with testosterone-estrogen combination therapy. Older cats suffering from hypothyroidism are aided by thyroid hormone replacement.

Accommodating the Older Cat

Sometimes, the little things make life easier—things such as multiple litter boxes with lower sides for cats with arthritis, a cold-air humidifier for cats that sleep under a drying furnace during winter months, or a heated blanket or hot water bottle for cats sensitive to cold.

Aged animals are more affected by the physical and mental ravages of stress. Providing a refuge for the cat disturbed by noise and the visits of the grandkids or other relatives or having a trusted friend or cat sitter care for your aged cat in his own environment while you are on vacation may be helpful.

Table 29: Body Changes Associated with Aging

Senses:

Hearing: Earwax production decreases. Eardrum becomes hardened and less elastic. Loss of hearing in higher frequencies occurs.

Sight: Tear production decreases. The lens thickens and becomes cloudy. Visual acuity decreases.

Taste: Taste buds are lost and taste sensitivity decreases. It may take more and more food to satisfy because taste perception is diminished. Saliva production declines, producing dry mouth.

Smell: diminished

Skin and hair: Loss of elasticity of skin. Skin becomes thickened. Hair becomes thin and brittle. Sebaceous glands located around face, flank, and tail decrease production of sebum and increase production of waxy secretions, which do not cover the hair, leading to dryness of hair coat. Nails become more brittle and footpads become thicker.

Bones, joints, and muscles: Bones become thinner and more brittle, increasing susceptibility to fractures. Cartilage hardens, elasticity decreases, and joint stiffness increases. Loss of muscle mass and tone.

Blood: Bone marrow, which produces components of blood, does not respond to the stress of blood loss or infection as rapidly as in young cats.

Immunity: Immunity decreases, leading to increased susceptibility to disease and cancer.

Neurology: Reflexes slow. The number of cells in brain, nerves, and spinal cord decreases, leading to reduced reaction to stimuli and partial loss of memory and senses.

Respiration: Vital lung capacity and efficiency decrease. The ability to clear mucus declines. Susceptibility to respiratory diseases increases.

Heart: Decreased cardiac output and thickening of heart valves may lead to congestive heart failure.

Urination: Ability of kidneys to filter decreases, leading to increased drinking and urination, and possible incontinence. Chemicals such as drugs may not be excreted effectively; reduced dosage of certain medications is indicated.

Reproduction: Irregular heat cycles, decreased conception rates, and increased fetal mortality. Increased incidence of tumors of reproductive tract in both males and females.

Digestion: Decreased liver function, intestinal absorption, and colon motility. Increase in constipation and gas.

Oral cavity: Inflammation and infection of gums and teeth, leading to bad breath and loss of teeth. Increased incidence of oral ulcers.

Other: Decreased sensitivity to thirst may lead to dehydration.
Decrease in thermoregulation may lead to cold and/or heat intolerance.
Decrease in amount or depth of sleep may increase restlessness and irritability.
Decrease in activity and metabolism, leading to reduction in caloric needs and often obesity.

Physical Conditions Associated with Aging

Constipation and Gas

There is an increased tendency for older cats to suffer from constipation owing to decreased liver function, intestinal absorption, and motility of the large bowel. Cats that are inactive and obese are especially prone to constipation, as are those consuming bones or the skeletons of insects and prey or those suffering from hair balls.

Chronic constipation can be treated by administering dietary fiber. Fiber absorbs water and increases fecal bulk. This increased bulk stimulates the defecation reflex and shortens the time undigested food spends in the intestinal tract, producing a softer, bulkier stool that is easier for the cat to pass. Fiber can be added in the form of high-fiber prescription diets or fiber-containing capsules, both of which are available at veterinary hospitals, or in the form of oat bran to a regular canned diet at the rate of one or two teaspoonfuls per can. Exercise approximately one hour after eating helps to stimulate elimination. Stool softeners and paste laxatives used for hair balls may be helpful, and as a last resort a soapy-water enema can be administered (do not use Fleet enemas on cats).

There is always a catch-22, it seems, and feeding high fiber is one of them. High-fiber diets are indicated for constipation, obesity, certain cases of diarrhea, hair balls, and treatment of diabetes. High-fiber diets can, however, increase the tendency to develop gas. Thankfully, cats are not as prone to objectionable odors from intestinal gas as are certain types of dogs, such as boxers and bulldogs.

Cutting back on the fiber in the diet in cases of gas is helpful. It is also important to provide a stress-free, quiet environment for eating and to offer the cat meals often, so that he won't bolt his food, ingesting air at the same time. Do not feed vegetables, milk, table scraps, or diets high in soy or wheat products. Do not exercise or excite the cat immediately before feeding.

I have noticed recently several new products to prevent gas in both pets and people. One product advertised for cats is CurTail. CurTail drops, which contain food enzymes, are administered to cat food containing soy or whole grains as an aid to digestion. For more information about CurTail, call 1-800-257-8650.

Cataracts

Senile cataracts, seen in very aged cats, are manifested by a cloudy or bluish appearance of the lens of the eye. In most cases, vision is not impaired to the point that treatment is necessary.

A common cause of cataracts in outdoor cats is fighting, which produces tearing and injury to the ocular lens by a sharp cat claw. Diabetes can also produce cataracts in cats.

In cases where the cat's vision is severely compromised by cataracts, regardless of cause, removal of the lens is curative. This is the day of specialists, and veterinary ophthalmologists work wonders by providing services such as cataract removal and placement of glass eyes in cats and other animals suffering from ocular trauma or disease.

Diabetes

Diabetes mellitus is reported in dogs, cats, horses, monkeys, ferrets, rodents, fish, and birds. In cats, aged and overweight individuals most often suffer from the disease.

Diabetes occurs when the insulin-producing cells in the pancreas fail to produce enough insulin or the insulin produced fails to do its job of allowing blood glucose (sugar) to pass into the cells of the body to provide energy for metabolism.

The signs of diabetes in cats and people are similar. There is an increase in urine production and urination because glucose filters from the blood into the urine. This increase in urination is followed by increased thirst and drinking to try to compensate for the loss of water in the urine. The cat's frequent trips to the litter box to urinate and to the water bowl to drink are usually the first clues that the cat has diabetes. The urine of untreated diabetics is sticky and sugary from the glucose it contains. One astute owner accurately suspected diabetes when she noticed a recurrence of ants in the cat's litter box.

Diabetes is confirmed by high levels of glucose in the blood and urine. Stress, by the way, will often cause a cat's blood glucose to be high, and repeated blood tests may be necessary to confirm that diabetes is causing blood glucose to be persistently high.

Primary diabetes is believed to have an autoimmune origin in older cats. Secondary diabetes can occur in response to long-term and high doses of certain medications such as corticosteroids and progestins. In the latter case, the diabetic condition is often resolved when the drugs are tapered off. About 10 percent of cats with clinical diabetes recover spontaneously. The remaining cats must be treated with insulin injections and/or with diet, depending upon the severity of the condition.

The goal of dietary treatment is to reduce or control obesity and to minimize high blood glucose following meals. This is usually accomplished with a high complex carbohydrate/moderate fat diet, usually a high-fiber diet. Frequent small meals are preferred to once- or twice-a-day feeding.

Some cases of feline diabetes are diagnosed when the cat is already severely ill and showing signs of dehydration and toxic ketones in the blood owing to fat breakdown. These cases must be hospitalized and treated as an emergency. In less severe cases, the veterinarian may elect to institute insulin therapy at home. The cat's caretaker will be instructed about giving insulin injections, feeding, monitoring glucose levels in the urine, and treating the hypoglycemia (low blood sugar) that can result when too much insulin is administered.

There seems to be good news for the treatment of diabetes in both humans and animals. Transplants of insulin-producing cells from the pancreas, already successful for treating diabetes experimentally, will hopefully soon become a reality for treating clinical patients.

Thyroid Disease

Both hypo- (underreacting) and hyper- (overreacting) thyroid glands are reported in greater frequency in older animals. See Chapter 10, pages 193–94, for more information.

Potassium Deficiency

Potassium is an electrolyte (salt) necessary for the proper functioning of the body's cells. Depletion of this electrolyte is seen frequently in cats with chronic kidney failure, liver dysfunction, diabetes, and lower urinary tract diseases, conditions seen often in elderly cats.

The signs of long-term potassium depletion are similar to those seen with normal aging—weight loss, anemia, thin hair coat, and listlessness. The difference is, however, that cats receiving supplemental potassium show a dramatic improvement in these symptoms.

The necessity for oral potassium medication and supplementation is based on clinical signs and laboratory tests. Concurrent diseases must be treated as well.

Cancer

Alexander was a lean and lanky red male cat that belonged to a writer and his wife. I loved to visit Alexander's house, for I always left with one or two autographed copies of his master's many published books and renewed hope for my own writing career.

Alexander, at age eighteen, developed a tumor in his mouth which I removed and sent to the veterinary laboratory. The pathologist there diagnosed an invasive form of cancer that was likely to recur, and so it did. I removed the mass each time it grew large enough to interfere with Alexander's eating, until finally, a year after the initial diagnosis, Alexander's owners decided the spunky cat had had enough and elected to have him put to sleep.

Tumors in cats, unlike those in dogs, are more often malignant (spread to other parts of the body) or invasive to surrounding tissue than benign (nonspreading) or noninvasive. Malignant tumors make up 75 to 80 percent of all feline tumors. Therefore, even the smallest bump, lump, or skin change on your cat should be examined by a veterinarian and perhaps biopsied for a pathology report.

As in humans, early detection of cancers in cats can mean the difference between successful treatment and death. A BB-sized mass in a cat's mammary gland may be the first sign of breast cancer. Redness and ulceration of the tip of the ear may be an indication of skin cancer. Chronic nosebleeds, tearing, and facial deformity may be a clue for nasal cancer.

Table 30: Warning Signs of Cancer

Abnormal swellings.

Sores that don't heal.

Bleeding from the mouth, nose, urinary tract, vagina, or intestinal tract.

Bad odor, especially from mouth and anus.

Difficulty eating, swallowing, breathing, or urinating.

Loss of appetite, weight, or interest in activity.

None of these signs are specific for cancer. They can have many different causations. One or more of the symptoms, however, should prompt you to consult a veterinarian about your cat's condition.

Cancer is primarily a disease of older cats. Over a lifetime, inflammatory events cause increased cell division and mutation. Some genetic damage causes cell death and injury, contributing to the aging process and increasing a cat's susceptibility to cancer. Cancer occurs when the body fails to mount an immune response sufficient to eliminate mutated cancer cells. Because the immune response to these cancer cells peaks at puberty and gradually decreases with age, most tumors occur in older cats.

The most commonly occurring cancers in cats are those affecting the blood, lymphatic system, and bone marrow (leukemias and lymphomas primarily). These cancers occur more commonly in cats infected with feline leukemia virus (FeLV); thus vaccination against FeLV is helpful in preventing these cancers, which constitute one-third to one-half of all feline cancer. Cancer of the skin is second in frequency of occurrence in cats, and mammary tumors are third.

The same conditions that predispose humans to cancer affect cats. Like people who develop lung cancer, cats living in a polluted environment suffer more instances of respiratory cancer. Overexposure to ultraviolet light can have the same result for lightly pigmented cats as for fair-skinned sunbathers—skin cancer. Breast cancer in cats closely mimics that in women and is hormonally related. Intact female cats are seven times more at risk for developing cancer of the breasts than those that have been spayed. Similarly, women who have had an early or artificial menopause have a lower risk for developing breast cancer.

Breeding and coloration are predisposing factors for some types of cancer. Cats with white fur on the face and ears are understandably more prone to skin cancers. Siamese seem to be relatively resistant to skin cancer, but are, unfortunately, very susceptible to breast cancer and lymphoma.

Animals with cancer receive the same type of treatment as human patients—chemotherapy, radiation, and surgery. Drugs used in chemotherapy are toxic to rapidly dividing cells. Since the cells lining the digestive tract are dividing rapidly, chemotherapy often results in vomiting, diarrhea, and reduced absorption of nutrients from the small intestine. The bone marrow, which produces white and red blood cells, is another site of rapidly dividing cells; thus chemotherapy often produces anemia and increased susceptibility to infection.

Radiation destroys rapidly growing cancer cells, but it also destroys normal cells adjacent to cancerous tissue. Radiation to the head or neck injures the mucous membranes, making mouth sores and ulcers a chronic problem. Sore throat and reduced saliva production may make swallowing more difficult. The sense of smell or taste may be altered. Radiation of abdominal and pelvic regions may lead to anorexia, nausea, vomiting, diarrhea, lactose intolerance, and malabsorption.

Surgery to remove cancerous tissue can also result in complications, depending upon the surgical site.

When your cat is recovering from cancer treatment, good nursing, nutrition, and loving care and concern are factors in his recovery. It has been proven statistically that people with life-threatening illnesses who have a close and intimate relationship with a pet have a better recovery rate than those without pets. Surely, the same is true of cats—those with a loving caretaker recover faster than those without.

Table 31: Nutritional Support for Side Effects of Cancer Treatment

Diarrhea or vomiting: low fiber, bland diet; vitamins A, B complex, C; frequent small meals; cooked starch only; protein supplements of soft-boiled eggs, cottage cheese, and finely ground, cooked meat.

Inadequate fat digestion, evidenced by fat in stool: low-fat diet.

Dry mouth: extra liquids.

Lactose intolerance: no milk products.

Anemia: vitamins, minerals, protein supplements, liver.

Low white blood cell count: sterilized food.

Anorexia: high-calorie nutrient supplements supplied in pastes, liquid, and syringe forms (Nutrical, CliniCare, Prescription Diet a/d); peanut butter; corn oil; powdered eggs.

Question

Dear Dr. Whiteley,

I have an old cat, age seventeen, that has been with me since I graduated from college. Miss Priss has seen me through two moves across state lines, the end of two romances, and the death of my parents.

My veterinarian recently told me that Miss Priss's kidneys are failing and that she has some other syndrome that I can't pronounce, much less spell. Anyway, I live in a small town with only one animal doctor, and although I trust the vet, I would like to get a second opinion about her treatment and about her chances for living her last few years in relatively good health.

Is there someone I can call? I am located a long way from a large city offering small animal vets and specialists.

Miss Priss's Mom in Prairieville

Dear Mom,

I recommend that you call the Dr. Louis J. Camuti Memorial Feline Consultation and Diagnostic Service sponsored by the Cornell Feline Health Center. The telephone number is 1-800-KITTY-DR (1-800-548-8937). Calls are taken weekdays 9 a.m. to noon and 2 p.m. to 4 p.m. EST. The fee is $25 for cat owners, payable to MasterCard, VISA, or Discover cards.

This service, open to both veterinarians and cat owners, offers consultations with veterinarians knowledgeable about feline diseases. You might have your veterinarian make the call for you since he is familiar with the case, or you might ask for copies of Miss Kitty's records so you can discuss the clinical signs, laboratory test results, and spelling of the syndrome you mentioned. Perhaps the best would be for you to set up a conference call at everyone's convenience.

I congratulate you for your pursuit of additional information that will help you care for Miss Kitty. A second opinion is an owner's right. Most veterinarians welcome the opportunity to discuss cases with other experts. As the old adage goes, Two heads are better than one.

H.E.W.

14

Saying Goodbye

Most of us outlive our cats and will experience the death of one or many cats during our life span. The bond between caretaker and cat is the primary factor in how we view and experience the loss or death of a feline companion. There are other factors, including our view of death and life after death for ourselves and cats, our relationships with other pets and people, and the degree of stress that we are experiencing now or in the recent past.

The Bond

Close bonding between people and cats, mentioned briefly in Chapter 1, makes saying goodbye difficult and heart-wrenching.

Cats We Identify With

Duke T. Cat was one of those tiger-striped, all-American felines with which everyone identifies. He was a working stiff—public relations and advertising executive for All Star Sheet Metal and Roofing Company in Amarillo, Texas. I didn't know Duke personally, for I came on the scene as veterinarian to All Star cats only after his death. I had noticed, however, Duke's mug on All Star's ads in my local newspaper.

When I first visited All-Star to patch up a stray kitty that good-hearted All Star employees had befriended, I noticed immediately the Duke photos, plaques, albums, and memorabilia that lined the outer office.

An All Star employee had rescued Duke from the middle of a city street when the little fellow was only a few weeks old. "He was the ugliest critter any of us had ever seen," Martha Woodard, secretary and treasurer of All Star, said, "but he was determined to stay with us." He grew up in the business and greeted employees and customers alike. Everyone loved and talked to and about Duke. He was an icebreaker and public relations guru. Soon Duke was company ambassador, appearing on All Star ads and contributing to local charities in the company's name.

As a member of the labor force, Duke allowed others to take themselves and their jobs a little less seriously. The working cat received a commendation from the Chamber of Commerce for his dedicated work in the community, and applied for and was granted a nonnegotiable workman's compensation check for on-the-job injuries suffered at the paws of an overzealous guard dog. As the years passed, letters, postcards, Christmas cards, telephone calls, and gifts poured in from people who felt a special bond with Duke. When Duke was killed in front of his business by a passing car, All Star employees and readers of Duke's ads and antics grieved.

I wonder what it was about Duke that made him loved by so many. He reminded me of C.K. because both were tiger-striped males. Apparently, he reminded other people of long-ago but beloved felines, for letters expressed such comparisons. Maybe it was his benign and gentle nature, captured so well in the All Star ads. Maybe he was plain and country, as the country-and-western song popular in my part of west Texas goes, "when country wasn't cool."

Joel S. Savishinsky writes in "Pet Ideas: The Domestication of Animals, Human Behavior, and Human Emotions," a chapter in *New Perspectives on Our Lives with Companion Animals:* "What dogs and cats are like as animals is reflected in what dog owners and cat-lovers are like as people." Whatever, Duke evoked identification and bonding, and when he died many said, "Goodbye, old friend, you will be sorely missed."

All Star, by the way, after a decent period of mourning, had a Duke look-alike contest and selected a young feline to continue their ad tradition. It didn't work, however, because the public would not accept a Duke replacement. The bond was with Duke, not just any tiger-striped cat. Replacing Duke would be like replacing George Burns with a cigar-smoking, wise-cracking younger man. It just can't be done.

The Feline "Child"

The bond between people and cats often expresses the nurturing parent/dependent child relationship. Savishinsky writes further:

> The pets who are child substitutes not only stand in for the children we do not have, they also stand—as our own children do—for the children we ourselves once were. By parenting the pets who represent us as we once were, we relive our own childhoods. The pet is simultaneously an animal, a child, and our own infantile selves.

The loss of a child, regardless of form, is a painful experience. The loss of our own childhood and grief from unresolved early issues may also be tied to the bereavement we experience when a child-substitute dies.

The Playmate

A youngster may relate to a cat as a companion or playmate. In the case of children raised in dysfunctional families, the family pet may be the only individual with which the youngster can communicate and receive acceptance. The loss of a beloved kitty may also be the first experience of death and bereavement that the child experiences.

The Cat Representative

Recall from Chapter 1 the man who established an intense bond with a cat belonging to his dead son and who experienced deep grief when the cat died (see page 14). Bonds between cats and people often have a way of establishing themselves around another person or situation.

A widow or widower may establish a close bond with the beloved cat of a deceased spouse; a pet born when our children were young may represent our early lives as a family. We grieve all over again when the cat dies, not only for the pet but also for our life with a beloved spouse or young children.

The feelings that surface when a cat develops a fatal illness of the same nature as the caretaker or another member of the caretaker's family may become intense, as the cat's impending death brings out the grief associated with the person's impending death. It may be easier to focus on the cat than the human, for we are still a nation uncomfortable with human mortality.

Cats serve as an outlet for dealing with intense feelings. Ann Ottney Cain, author of "A Study of Pets in the Family System," in *New Perspectives on Our Lives with Companion Animals,* suggests that some people focus their attention and emotions on pets as a way of avoiding interaction with their human family. Several people responded to a questionnaire in Cain's study with examples of situations: The husband relates lovingly to the pet, but the wife feels left out; mother talks to the cat, telling the feline companion things she wishes to relay indirectly to the child; husband and wife both offer affection to the cat, thus avoiding the giving of affection to each other; family members focus on the pet rather than communicating with or confronting each other.

When the cat dies, the family dynamics shift, often uncomfortably. One family member may be glad the cat is gone and feel guilty; another may mourn the loss of the object of his affections with intense sadness.

Regardless of the psychological reason for the bond, the greater the emotional investment in the cat, the greater the feelings of loss and grief when the relationship ends.

View of Death

How we view death is crucial to letting go of those that die. If death is the end of all existence, dying brings with it a finality that is hard to bear. If we believe, as modern physicists are discovering, that matter (the physical body) is also energy and that energy is never lost but only changes in form, then dying is the transition into another form, spiritual in nature.

Do pets have souls? I believe that all living things have a soul, a spiritual energy, that is eternal. How that soul is manifested and where it goes upon death is part of the mystery that makes life and death such an exciting adventure.

Interestingly, some people who describe near-death experiences report that they are met after death, at the end of the tunnel of light, by deceased pets. In an article titled "Do Animals Survive Death?" appearing in *Veterinary Forum,* Scott S. Smith cites anecdotal evidence of life after death for animals. He includes a report from a veterinary technician who saw the following after the euthanasia of a dog: "Suddenly, he saw a 'luminous haze come out of the body. It clung to it for a few moments and then became a cloud over the old body, after a minute disappearing into the floor.' " Others reported seeing the "ghosts" of deceased pets.

Do animals go to heaven? It depends upon one's view of heaven, of course. I believe that love, which is a form of energy, is never lost, so if heaven is a place of love then C.K. will surely be there. If heaven is a concept outside our limited view of time and space, perhaps all energy becomes the eternal oneness that is heaven, God, the happy hunting ground, or the universe, depending upon one's chosen terminology.

The purpose of this section is not to foist my spiritual beliefs upon you but to suggest that it is appropriate to believe what you believe about the spiritual survival of your cats after death. It is comforting to me to believe that the love I shared with C.K. is eternal and everlasting.

Relationships with Others

In a study of bereavement counseling for grieving pet owners conducted at the Veterinary Hospital of the University of Pennsylvania in 1980–81, certain indicators of bereavement emerged. Cat owners who have the most difficulty recovering after the loss of a pet are those who have lost an only-pet with which they have spent over seven years and those who live alone or with a spouse only. People who have multiple pets or those with an extended or nuclear family including children seem to bounce back sooner after the death of a pet. Significant others serve as an in-house support group, and carrying on for the sake of remaining pets and/or children may help direct our focus away from our loss.

The University of Pennsylvania study showed that cat owners are more likely to be referred for bereavement counseling than dog owners. Aside from cat owners who keep cats as status symbols or mouse catchers, we cat owners bond closely with our feline companions and grieve intensely at their loss from euthanasia, disease, or injury.

Women are much more likely than men to seek or accept bereavement counseling. In spite of the recent men's movement, men are still denying and hiding their emotions and feelings. I believe that macho men are often the ones who suffer the most when a beloved cat dies; however, they are essentially isolated in their grief. A man with no wife or girlfriend on whom to vent his emotions is reluctant to ask for support from his buddies at the Truck Stop Cafe or Rifle Club when Fluffy dies.

For the most part, the elderly are hit the hardest when a beloved pet dies. This group of our population also suffers more from isolation and loss of relationships and support, and in many cases, from reduced mobility and a lifestyle offering fewer diversions and responsibilities.

Life Stress

This is the old last straw syndrome. When too many stresses occur in close proximity, our ability to deal with loss diminishes. The Pennsylvania study revealed that people who have problems dealing with the death of a pet are those who have suffered multiple stresses during the last year: loss of job, death of a significant other, relocation, etc. Stress-related illnesses occur in greater frequency in people who have high scores on life-stress tables.

Table 32: Life-stress Factors

Death of spouse or family member.

Divorce or separation.

Serious illness or injury to self or family member.

Death of pet.

Reversal of financial situation.

Loss of job, change in job, or retirement.

Birth of child or children leaving home.

Marriage.

Change in residence.

Death of close friend.

The Grieving Process

Any loss from the smallest (such as the loss of an article of clothing) to the largest (those mentioned in the life-stress table) must be grieved. Losses accumulate, and if not processed and accepted, lead to depression.

Unresolved issues have a way of returning to attention in order, I think, to afford us another opportunity to lay them to rest. Sometimes, our dying cat reminds us of another event. The feelings are familiar, and if we allow ourselves a moment of reflection, we might discover that we feel guilty and vulnerable, just as we were at ten years old and the first Fluffy died. Then, another thought surfaces and brings with it the realization that the loss of that first pet coincided with our parents' divorce. All we can feel now is loss, and it is awful. We cry and cry for our cat, but know that all of our losses are somehow tied together, and it feels overwhelming. This is not the time to deny or push away the pain. Now is the time to grieve, cry, scream, feel the anger, resolve guilt, and forgive ourselves. Grieving is healing.

Stages of Grief

Dr. Elisabeth Kübler-Ross, who now focuses on people with AIDS, was the first to study the dying process and to develop a guideline for the stages that people go through in coming to terms with death, the ultimate loss. Kübler-Ross coined the stages of grief: denial, anger, bargaining, depression, and, finally, acceptance.

Although the stages are set down in a stepwise fashion, one's grief is personal and individual in nature. The grieving process may last only minutes for small losses or forever for life-shattering ones. One may think he has finished the whole process with the sentiment, "Thank God, I managed to get through that," only to find the old grief resurfacing again and again until it is truly over and completed. You may bypass a stage and return to it later or may vacillate between various stages.

Denial occurs when something is too overwhelming to accept. It lasts until we are able to mobilize our coping resources. This early stage of grief can manifest itself as a refusal to believe whatever has occurred. The veterinarian has told us that Fluffy has died, but it doesn't register.

I know a veterinarian who had to tell a physician that his pet had been dead for several hours. Of course, on one level the medical doctor "knew," but the loss was just too terrible to comprehend at the moment.

The man and his dog had been on a vacation together in the mountains, and the dog had fallen over a rocky embankment to his death below. The physician had climbed down a cliff to rescue his dog, walked many miles back to his car carrying his pet, and then driven many more miles to find a veterinarian. On the way to the vet's office, he may have experienced feelings associated with grief such as anger (Why don't those lazy good-for-nothing veterinarians work on Saturday afternoons?), blame (If my dog dies, it's my wife's fault; if she had come with me as I wanted, I wouldn't have brought the dog), bargaining (If my dog will just be all right, I'll stay at home with him), guilt (It's my fault; if I hadn't brought the dog with me, this wouldn't have happened), or regret (Why didn't I leave Rover at the kennel?). A few days after the accident, the physician may have felt depressed. He had to return to his home and work, and explain the loss of his dog to others. Denial, anger, bargaining, guilt, and regret had become too much of a struggle, and he was now overwhelmed with his loss. He felt the pain and sadness of his loss. In a few weeks, the man realized that he was not focusing so much attention on his dog. He had begun accepting the dog's death and had even caught himself looking at puppies in the window of the pet store at the mall.

Of course, I don't know what the physician was feeling, but denying, feeling angry, blaming, regretting, bargaining, feeling guilty and depressed, and accepting are all part of the grieving process. When we feel these emotions we are not the only ones who have ever felt this way after a loss, and it may be helpful to accept ourselves and our common human reactions. Grieving is a normal process of living, and physical or mental illness may be the penalty for unresolved grief.

Unresolved Grief

I know a woman who took her critically ill cat to five veterinarians working in different hospitals, hoping for a cure. When the cat died the woman picked one of the veterinarians to blame. I have known and worked with this veterinarian for several years; his sense of caring and responsibility, and his expertise are beyond reproach. He, like all of us, is capable of poor judgment or human error, but in this particular case he had seen the cat only briefly to confirm the previous veterinarians' diagnosis. The woman had then taken the cat to the next cat doctor.

The woman came into the veterinarian's office while I was working there in his absence and berated him in front of all the clients in the waiting room. I showed the woman into an examination room, and she proceeded to tell me that Dr. X had killed her cat. Several years later she still calls periodically to yell at poor Dr. X for the death of her cat. This woman is full of anger and blame. This is the grieving process gone awry, as is the case of the owner suffering prolonged depression, guilt, or codependency involving other pets. When the pain of living with unresolved grief becomes too great, the affected person will begin the process of healing by seeking professional help or by finally feeling the emotions necessary for acceptance and healing.

Resolving Grief

SUPPORT–One of the terrible things about the death of a pet is that other people minimize our loss. Often, we feel so alone in our grief. We fear that if we express our feelings of sadness and depression, others will say, "Oh, you'll get over it. After all, it was only a cat." In some cases, our feelings of loss are as great as or greater than those at the death of a human loved one, yet we are unable to express how bad we really feel. We may try to deny our feelings in an effort to conform to what others expect, thus delaying our recovery.

The first step is to give ourselves support and to seek out friends who will allow us to vent our feelings without judging us. It may be worthwhile to seek out a pet-loss hotline, support group, or individual counselor. We may need to take a few personal days off from work. We definitely need to be kind, gentle, and understanding with ourselves.

When C.K. died, bargaining, denial, guilt, depression, and finally acceptance were the emotions I could identify. I felt very much responsible for his death. I had gone to a veterinary meeting one night in my usual mad rush against time, and C.K. followed me out the front door. He normally spent a few minutes outside and then wanted to come back inside. I held open the door and admonished him to go into the house. When he didn't immediately respond I shut the door, leaving him outside until my return from the meeting. When I got home I called and called, but C.K. didn't come. In another hour I called for him again. Finally, at midnight, I got up and searched the alley with a flashlight, but still couldn't find him. *Just let him be okay,* I prayed, *and I'll always take care to put him inside in the future.* I spent a sleepless night, chastising myself for leaving him outside. As the first light of morning peeked over the trees, I looked out my front door, and there he was under the rosebush. I knew immediately that he was dead. His head was smashed on one side, and I could picture him running across the street for his home after being hit by a passing car. *Oh, how could I have been so careless!* I kept repeating in my mind. *If only, if only . . .*

I canceled my appointments, asked another veterinarian to take my emergency calls, and went to bed and cried for two days. For months, I continued my habit of nudging the edge of my bed where C.K. slept and was always surprised to find that his furry body was absent.

My family, friends, and clients shared my grief, and offered the most precious of commodities—acceptance and understanding. Several friends and clients sent cards. Later, one of my clients asked my husband for a photo of C.K., which she used as a model to have a pet portrait made for me as a surprise gift. I have C.K.'s painting hanging in my office. He has been dead over four years now, and I miss him still.

Table 33: Pet-loss Resources

Pet-loss Support Hot Lines

Companion Animal Association of Arizona at Phoenix: 24-hour hot line staffed by trained volunteers: (602) 995-5885.

Grief counseling by University of California at Davis veterinary student volunteers; staffed weekdays, 6:30 to 9:30 P.M. Pacific time. (You must call in; volunteers cannot return long-distance calls.) (916) 752-4200.

Grief counseling by University of Florida at Gainesville veterinary student volunteers; staffed weekdays, 7 to 9 P.M. Eastern time. (904) 338-2032.

Counseling and Support Groups

The Animal Medical Center of New York: on-staff social workers for counseling and bereavement groups. (212) 838-8100.

Most colleges of veterinary medicine offer grief counseling for pet owners dealing with the grief and anxiety of a pet's illness and/or death.

The Delta Society, an organization devoted to study the human/companion animal bond, offers resource material and a listing by state of organizations and individuals offering hot lines, support groups, and counseling dealing with pet loss. (206) 226-7357.

Euthanasia

"Putting an animal to sleep" is, in my way of thinking, one of the kindest decisions a pet caretaker can make if his pet is experiencing minimal quality of life, requires extensive life support that the owner cannot offer, or is a danger to others. The option of euthanasia can be abused, of course, but I have found that owners often wait too long rather than act too quickly to elect euthanasia.

Deciding the "if and when" is a personal decision. I was guilty of silently judging an owner's decision to prolong a cancer-ridden cat's life when I should have been vocally supportive. Later, after the cat had died at home, I learned that this woman was the mother of a child suffering from the same cancer as the cat. How could I ask that mother to give up hope for her cat or for her daughter? Refraining from judgment has something to do with walking in another's moccasins. For this reason I try never to make this decision for someone else.

Euthanasia should be a choice made jointly by all members of the cat caretaking group. Your veterinarian may be able to guide you by offering a prognosis (a prediction of the future course of illness or behavioral problem). All options should be considered. Guilt about a euthanasia decision, for or against, serves no useful purpose, and talking the decision over with personal and professional support groups helps us let go of guilt.

Once the decision is made to put the cat to sleep, the time and place can be selected, the grieving process begun, and plans for burial, cremation, or other means of honoring the pet put into action.

The veterinarian will ask you to sign a form that gives legal permission for him or her to perform euthanasia. It states that the animal has not bitten anyone within the last fourteen days, in which case he should be quarantined for rabies before euthanasia.

The current method of euthanasia used by veterinarians is an injection into the pet's vein of a concentrated drug used for anesthesia. The animal literally goes from sleep to death. I like to administer a tranquilizer before the IV to relieve any anxiety that may be associated with the injection itself.

Most veterinarians will permit you to be present at the time of euthanasia. This, again, is an individual choice. If your veterinarian offers home service, euthanasia can be performed in the cat's home environment. If this option is elected, make sure that other pets are not in the immediate area; animals are sensitive to the intense feelings and to the dying act. Whatever the situation, you may ask to spend a few minutes alone with your pet to say goodbye prior to and/or after euthanasia.

Your veterinarian can take care of the cat's body, or you may make your own arrangements. This should be decided before making the appointment. It may be a good idea to ask a friend to be with you or to drive you home or offer other support. And yes, it is okay to cry.

Honoring

When I moved to a house in a rural community, one of the first things I found when examining my new property was a small grave marked with stones, a wooden cross, and green plants, scarce here in west Texas. The family who lived here before me had left a memorial to a beloved pet.

For some, it is fitting to leave Fluffy at the veterinarian's office and remember the good times. For others, a burial or cremation or even a taxidermy or freeze-drying service is needed. Communication and sharing are healing for some, and they write poems, stories, and books about beloved cats. Others volunteer at a humane shelter or pet-visitation program, fund a scholarship or wing of a veterinary hospital, donate to favorite animal-related charities, or plant a tree in memory of their beloved pet.

At least one newspaper—*The West Chester County Daily Local News* of West Chester, Pennsylvania—publishes a pet obituary called "Pause to Remember." What a nice way to honor pets and grieving owners.

Burial/Cremation

People from many cultures have derived comfort from the burial or cremation ceremony: Ancient Egyptians paid careful tribute to their deceased cats by embalming them, wrapping them in linen, and placing them in a casket or sarcophagus.

Burials can take place in the backyard or back pasture or in a pet cemetery, depending upon your preferences or needs, location and condition of the ground, and city ordinances. Costs for a pet service and burial range from essentially nothing to thousands of dollars, if elaborate trappings such as a marble engraved stone and a stainless steel casket are wanted.

To locate pet cemeteries, ask your veterinarian, consult your yellow pages, or write the director of the International Association of Pet Cemeteries, P.O. Box 1346, South Bend, IN 46624, for a list of guidelines and location of members.

Many pet cemeteries require you to purchase a casket, urn, or marker from them. If you are performing your own burial or have made arrangements for cremation, you can make your own burial box or urn.

Our city animal-control facility offers owners of pets private cremations for a low cost of $25. Often I acted as funeral director by suggesting this option, helping pick out a pretty container for the ashes, and serving as go-between for the bereaved and animal-control personnel. It may have been the most helpful service that I offered my clients.

Whatever seems appropriate to you and your cat is best. If the cat loved the outdoors, burial or scattering of ashes at the base of his favorite tree might be fitting. If he loved to curl up in front of the fireplace, an urn located on the mantel or hearth might be better. If you opt for a pet cemetery, place something significant on the grave marker—his picture or a personal poem, for instance.

Funeral Services

Funerals were not created for the dead, for they are gone from this physical environment. Funerals are designed to give the living and the grieving comfort. In that wonderful movie about friends, *Fried Green Tomatoes,* there is a touching funeral conducted for a child's severed arm. Certainly a pet is a friend and also deserves a funeral.

The type of service that you create depends upon your beliefs and preferences. It may be as simple as reading a poem or placing a child's drawing of his pet upon the grave site or as elaborate as a service complete with clergy and special music.

Table 34: Inspiration for Services

Prayer of St. Francis of Assisi: Lord, make us instruments of thy peace. Where there is hatred, let us sow love; where there is injury, pardon; where there is doubt, faith; where there is despair, hope; where there is darkness, light; and where there is sadness, let us sow joy.

Divine Master, grant that we may not so much seek to be consoled, as to console; to be understood, as to understand; to be loved, as to love; for it is in the giving that we receive; it is in pardoning that we are pardoned; and it is in the dying that we are born to Eternal Life. Amen.

Poem: All things bright and beautiful,
All creatures great and small,
All things wise and wonderful,
The Lord God made them all.
—Cecil Frances Alexander, 1818–95

Scripture: Praise the Lord from the earth,
Sea monsters and all ocean depths;
Lightning and hail, snow and clouds,
Strong winds that obey his command.
Praise him, hills and mountains,
Fruit trees and forests;
All animals, tame and wild,
Reptiles and birds.
Praise the Lord!
—Psalm 148:7–10

Hymn: For the beauty of the earth
Sing, O, sing always:
For the sun and for the moon
Sing God's praise today.
Lift your hearts and lift your hands
To Him who gives us breath;
Join the choir of creatures in this
Hymn of life o'er death.

. . .

For the pets we hold in trust,
Loyal friends and true;
For their lives and for their love,
Making each day new;
Take our promises, now given,
Lord to be always
Stewards of their welfare and
Friends for all their days.

. . .

In the Mystery of life,
Humbling every mind;
In the turn of birth and death,

Good Lord help us find;
All your love you offer to us,
Man and beast as well;
As partners on this pilgrimage, by
Love, we your Love tell.
—Words by Father Richard L. York; melody *Adoro Devote*

Benediction: Good Lord send us forth
To lovingly caress
Your great creation.

Helping Others Cope

Children

Among those who suffer most from the loss of a cat are children. The degree of suffering is influenced by the child's age and emotional development, the closeness of the bonding between child and pet, the relationship between child and parents, and the child's previous experience with grief and death.

Under age five, most children have a difficult time grasping the idea of death and separating it from dreams, fantasies, or just something they saw on television. These young children can become confused about the death, or they may even believe the pet died from something they did, such as pulling the cat's tail.

Honesty is almost always the best policy in helping children deal with the death of a pet. Specifically tell the child that the cat died. Do not use confusing words like "sleeping," "passed away," or "lost." Encourage the child to talk about his feelings. Share some of your own feelings of grief. Be patient if the child asks the same questions over and over. Explain what happened: Fluffy was very sick and died; Fluffy was hit by a car and his body was so broken that he died. Allow the child to take part in a funeral service. Encourage the child to color or draw a picture of Fluffy.

Older children, between five and twelve, usually have some experience with or understanding of death. They often seem to be overinterested in the "gory details" and in shocking adults by asking questions about the pet's injuries. These questions are the child's way of trying to understand and to come to grips with the reality of the death, and should be answered directly and honestly.

Teenagers can have the most difficulty accepting a cat's death or impending death, especially in the case of euthanasia. Rather than being sheltered from a euthanasia decision, the teenagers closely bonded with the cat should participate in the decision. Otherwise, they may feel their parents are responsible for the pet's death.

Table 35: References for Children and Those Who Feel Like Children

Death of a Pet: Answers to Questions for Children and Animal Lovers of All Ages. Guideline Publications, P.O. Box 245, Stamford, NY 12167 (1-800-552-1076).

The Dead Bird, by Margaret Wise Brown (New York: Dell) 1979.

The Tenth Good Thing About Barney, by Judith Viorst (New York: Macmillan) 1987.

The Fall of Freddie the Leaf, by Leo Buscaglia (New York: Holt, Rinehart and Winston) 1982.

Mustard, by Charlotte Graeber (New York: Macmillan) 1982.

I'll Always Love You, by Hans Wilhelm (New York: Crown) 1985.

The Accident, by Carol Carrick (Boston: Houghton Mifflin) 1981.

The Elderly

Dr. Leo Bustad, author of *Animals, Aging, and the Aged,* says, "The most frequent disease of old age is loneliness of spirit. A majority of the aged are sleeping their lives away, waiting to die, but with a pet they can be spared this loneliness."

Older people who have outlived relationships, both human and animal, may elect loneliness over fear—fear of the grief process should they invest their emotions in a pet that dies or fear for the pet's life and care if they should die first. Yet with assistance and support, many of these fears can be relieved.

Perhaps one of the kindest things that middle-aged children can do for aging parents is reassure them that beloved cats will be cared for in case of the parents' death or incapacitation. This concern plagues young, single adults as well, and instructions about care and disposition of cats should be included in an individual's will.

Many community programs help aging and handicapped individuals live more fulfilled lives with their feline companions. These services include pet visitations to nursing homes; emergency care for pets when owners are ill, are hospitalized, or die; and programs that help the elderly and handicapped with pet care. Check with your local humane societies for programs in your area.

Animal Grief

Does a cat grieve when his beloved human companion dies or is hospitalized for a lengthy period? What does he feel and think when a feline sibling of fifteen years goes away to the veterinary hospital and never returns?

When George O'Brien died suddenly and unexpectedly of a heart attack, Joie, Mousie, and Merri Anne, the feline members of George's family, grieved. "I know they miss him," George's daughter Cheryl wrote me, "plus, they must pick up the feelings of grief that Mother and I are experiencing at this time."

These cats reacted differently from each other, but all showed at least one symptom of stress—anorexia, depression, seeking seclusion. Fortunately, these cats had two loving caretakers remaining, Cheryl and Dorothy, and they eventually responded to the understanding and concern that the two women extended to them.

What can you do to help a feline say "goodbye"? Do as much as you can to eliminate other stresses from his life. Accumulated stress acts on cats just as on people: It makes each event more stressful. This is not the time to bring a new cat into the household, board the cat in strange surroundings, change the cat food, or leave the cat at the vet's for neutering.

If possible, spend time talking to and reassuring the cat. In some cases, providing familiar objects with the deceased's smell or playing ball with the cat just as Dad did is helpful. Provide a safe, quiet environment away from the friends and relatives who always gather at this time; offer nutritious food and plenty of fresh water. If a friend who is familiar to the cat asks to help, let him oversee the care of the cat during this stressful time for all household members.

Starting Over

Some people are ready to start over with a new cat immediately after the death of the previous one and some give up on cat caretaking completely.

Ann Spencer writes in *People, Animals, Environment: Bulletin of the Delta Society* about a study of bereaved cat owners and the cats they choose as replacements:

> The short-time replacers most likely lost their pet or it was killed in an accident, the animal was buried at home, and the replacement animal closely resembled the previous animal. Those pet owners who waited the longest to replace their pet had experienced the anxiety of a long illness and/or euthanasia, the pet was left with the veterinarian for disposal, and they selected a replacement pet completely different from their original animal.

However, Spencer concedes:

> Finally, there is at least one major factor which must be considered in evaluating these choices. That is the concept of bonding, which can completely alter choice. Prospective owners looking for a specific color, or an "anything but" request sometimes went away cuddling something completely different. The openly stated reason for this was instant bonding: "I know I said not black and white, but I felt this one come to me. As soon as I picked it up, I knew!"

And here we are back where we started—choosing cats, covered in Chapter 1.

Question

Dear Dr. Whiteley,

I am an aging widow living alone with two aging cats. These cats are my whole life, as I have outlived all my immediate family.

My primary fear in life is that I will become ill and be unable to care for myself or my cats. If I die or am forced into some kind of institution, what is to become of my wonderful babies?

Mrs. D. L. Brown
Waco, Texas

Dear Mrs. Brown,

I am pleased to inform you about the new Companion Animal Geriatric Center at Texas A&M University in College Station, not far from where you reside. This center is dedicated to caring for pets belonging to owners who are no longer able to care for their companion animals.

The center is located on a ten-acre complex on the veterinary school campus, and Texas A&M veterinarians will care for the pets in a homelike atmosphere while studying the aging process without experimentation.

Each animal resident of the program must be supported by a donation to the center. Contact the College of Veterinary Medicine, Texas A&M for more information: Texas A&M University, College of Veterinary Medicine, College Station, TX 77843-4461. Telephone: (409) 845-5051.

You may be able to take your feline companions with you if you move to certain types of public housing. Section 227 of the Housing and Urban Rural Recovery Act of 1983 provides that an owner or manager of federally assisted rental housing for the elderly or handicapped may not prohibit or prevent a tenant from owning or having common household pets in a dwelling unit. The Housing and Urban Development Office (HUD) can provide information on which units are covered under the law.

Best wishes to you and your feline companions for many more years of companionship.

H.E.W.

Index

abortion, 197, 208, 216, 228
Abyssinians, 21, 170, 213, 214
activity level, 9, 11, 13, 14, 235; of kitten, 22
age of cat, 22, 232–36; feline/human age comparison, 235–36
aggression, 13, 14, 24, 27, 48, 49, 50, 61, 74, 130, 139–49, 169, 170, 179, 183; between males, 146–47, 199, 202–208; maternal, 225–26; paternal, 226–27; play, 143–44, 147; redirected, 142–43, 147; sexual, 146, 198–200, 202–208; territorial, 145–46; toward other cats, 145–49; toward people, 140–45
aging. *See* elderly cat(s)
AIDS, 187, 254
air travel, 85, 88–90
allergies, 9, 11, 15–17, 157, 160, 161; food, 110, 157–61, 177–78, 245; solutions for, 15–17; and stress, 158, 160
Allerpet, 16
amino acids, 98
anemia, 185, 242, 245
animal control, 189–90
Animal Medical Center, New York, 81, 258
animal shelters, 12, 145, 206
anorexia, 105, 164–66, 176, 178, 188, 235, 245, 266
antibiotics, 116, 160–61
antifreeze poisoning, 179
anus, 31, 38; absence of, 214; grooming, 109
Ardrey, Robert, *The Territorial Imperative*, 74
arthritis, 132
aspirin, 178
asthma, 157, 158
attachment period, 27
attention-seeking cat, 152, 156, 169, 218
bacterial infections, 161
baldness, 108, 170, 213
bathing, 31, 63–64
bed(s), 109–10; for kittens, 217–23; wetting, 138–39, 157
behavioral problems, 9–10, 12, 129–53; aggression, 139–49; attention seeking, 152; correcting, 129–52; and declawing, 59–61; destructiveness, 149–52; diseases causing, 130, 132–33, 175–91; eating disorders, 100–108, 164–66; excessive sucking, 105–106; factors in, 130–31; grooming, 109; isolated kitten, 24–25; jumping on counters, 151–52; litter box, 37, 54, 56–57, 117–18, 130, 131–39, 151, 159, 237, 241; obsessive-compulsive disorders, 167–72, 178; scratching, 130, 140, 144, 147, 149–51; sexual, 130–31, 200–206; spraying, 134–39; stress-related, 156–73; veterinarians for, 152–53
belly, exposed, 48, 49, 71
benzoic acid, 178
bird hunting, 70, 71, 72, 96
birth-control pills, 209
birthing, 217–21
biting, 140, 144, 147; compulsive, 168–72, 178–79
bladder, 111; incontinence, 237; infection, 132, 133
blood, 179, 181, 238, 241, 244
boarding, 10, 90–91, 92–93
boat travel, 90
body language, 38, 46–49, 140, 146
bonding, 14–18, 22–23, 28–29, 224–25, 248–67
bones, of elderly cat, 238
Borchelt, Peter, 54
brain, 37, 42, 45, 110, 162, 172, 179, 183; disorders, 182, 183–84, 187, 188, 213

Braun, Lilian Jackson, 6
breast(s), 218; cancer, 206, 208, 243, 244; enlargement, 205; lactating, 96, 156, 217, 218, 221–23, 228–29
breathing, 67–68, 238, 244
breed(ing), 8, 11, 20–21, 212–14, 233; dating services, 212–13; inbreeding, 212, 235; and life span, 233, 235; management, 214; personality traits, 20–21, 212; predisposition to genetic and congenital disorders, 212–14; and pregnancy, 214–21; and season, 194–97, 199. *See also* sex; *specific breeds*
Bronson, Roderick, 233
brotherhood, 73–74
brushing, 9, 47–48, 61–63
bully cat, 144–45
burial, 260–61
Burket, Carmen, 12
Burmese cats, 21, 130, 145, 170, 213, 214
Bustad, Leo, *Animals, Aging, and the Aged,* 264
bus travel, 90

Cain, Ann Ottney, 250–51
calicivirus, 189
calico cats, 31, 167–68, 169, 204
Camuti, Louis, *All My Patients Are Under the Bed: Memoirs of a Cat Doctor,* 7
cancer, 132, 161, 163, 164, 185–87, 204, 206, 233, 238, 258; in elderly cats, 243–45; warning signs, 243–44
canned food, 99, 236
cannibalism, 226, 227
cardiopulmonary resuscitation, 67–68
carpets, urination on, 12, 55, 133, 137
carrier, cat, 30, 84–86, 88–89, 114–15
car travel, 27, 31, 83–88, 162
castration. *See* neutering
cataracts, 240–41
caterwauling, 200–201
catnip, 27, 55–56, 106, 178, 202
cemeteries, pet, 260, 261
cervix, 195, 201
cesarean section, 220
chasing, 25
chemotherapy, 244–45
children, 13, 14, 15; and aggressive cat, 141; allergic to cats, 15–17; cat as substitute for, 249–50; and death of cat, 250, 263–64; and kittens, 23, 24, 27, 29
choice of cats, 8–11, 20–23
city cats, 73, 114; and high-rise syndrome, 81–82. *See also* indoor cat(s)
claws, 57–61, 70; declawing, 16, 59–61, 72, 145, 150, 173; pedicure, 57–58, 149, 237; scratching post, 58–59, 149–50; scratching problems, 149–51; and territorial marking, 77, 138–39, 149, 199, 208
climbing stairs, 124–25
CliniCare, 165, 245
clumping litters, 54
codependency, 17–18
collar, 72, 83, 88
coloration, 21, 204, 213, 244
colostrum, 221
combing, 61–63
come (command), 118–19, 169
commands, teaching, 118–20
commode, 55–56; drinking from, 99
communication, 45–50
compulsive behaviors, 167–72, 178
congenital disorders, predisposition to, 212–14
constipation, 46, 239–40
coughing, stress-related, 157–58
counters, jumping on, 151–52
cremation, 260–61
cryptorchidism, 204
CurTail, 240

dander, 16
dating services, 212–13
deafness, 21, 39, 213
death of cat, 12, 67–68, 157, 177, 188, 232, 234, 235, 247–67; burial/cremation, 260–61; and children, 250, 263–64; and elderly, 264–65, 267; euthanasia, 252, 258–59, 264, 266; funerals, 261–63; grieving process, 12–13, 252, 253–66; kittens, 226–28; and replacement pet, 266; view of, 251–52
declawing, 16, 59–61, 72, 145, 150, 173
defecation, 117–18; constipation, 46, 239–40; diarrhea, 91, 99, 100, 110, 132–33, 159–61; in elderly cat, 239–40; housesoiling, 131–39; and kittens, 52–57, 222; and parasites, 183–84; potty training, 52–57; as territorial marking, 77, 131, 134, 138, 208. *See also* litter box
dehydration, 164, 165, 176, 239, 242
depression, 138–39, 265
despot cat, 148–49
destructiveness, 149–52
diabetes, 12, 54, 135, 209, 240, 241–42

diarrhea, 91, 99, 100, 110, 132–33, 157, 176, 179, 187, 219, 244, 245; stress-related, 159–61, 164
diet. *See* feeding
digestive disorders, 213, 214, 239–40; stress-related, 159–66
disease(s), 175–91, 233–34, 238; behavioral problems due to, 130, 132–33, 175–91; common signs of, 176; distemper, 161, 163, 189, 228; in elderly cat, 238–46; epilepsy, 130, 182–83; feline immunodeficiency virus, 161, 182, 187; feline infectious peritonitis, 161, 182, 188, 189, 228; feline leukemia, 132, 161, 163, 182, 183, 185–87, 189, 190–91, 214, 228, 244; and life span, 233, 234; poisoning, 106–107, 130, 161, 164, 178–80; rabies, 176–77, 189, 190; thyroid problems, 181–82; toxoplasmosis, 182, 183–85, 228; vaccination schedule, 189
displacement grooming, 109, 167–72
distemper, 161, 163, 189, 228
dog(s), 7, 12, 28, 43–44, 97, 98, 99, 100, 115, 140, 143, 158, 165, 176, 177, 182, 203, 209, 240, 243, 255; food, 98, 99; genitals, 195; living with cats, 9, 30–31, 148–49; obese, 102
dominance, 7–8, 75–77, 148–49, 161, 199, 202, 203, 204–205, 226–27
don, 73
door, cat, 79
doorbell, ringing, 121–22
dreams, 110
drinking, 99–100
drooling, 170
dry food, 37, 64, 96, 97, 99

ear(s), 34, 41, 167, 176; disorders, 21, 39, 213, 214, 238; grooming, 65; infections, 65, 162; of kittens, 221, 222; language, 47–49; mites, 65; wax, 65
eating disorders, 100–108, 164–66
eggs, fertilization of, 214–15
Egyptian cats, 6–7, 260
elderly cat(s), 12, 22, 54, 109, 231–46; body changes in, 238–39; cancer in, 243–45; care of, 236–39; death of, 247–67; elimination problems, 132, 237, 239, 241–42, 243; feeding, 236, 237, 239–43, 245; grooming, 58, 132, 237; obese, 101, 236, 239, 240–42; physical conditions in, 239–46; thyroid problems, 181, 237, 242
electrolytes, 165
embryos, 215–16
enzymes, 96, 100, 160, 214, 240
epilepsy, 130, 182–83
esophagus, dilated, 163
estrogen, 195, 205, 206, 208, 215, 224, 237
estrus. *See* heat
etiquette training, 52–68; administering medications, 65–67; grooming, 57–65; litter box, 52–57
euthanasia, 252, 258–59, 264, 266
exercise, 103, 105, 162, 236, 240; lack of, 234; training, 123–26
exposure to animals, 30–31
exposure to humans, 28–29
eye(s), 34–36, 40, 63, 146; cataracts, 240–41; grooming, 65; of kittens, 221, 222

fabric chewing, 107–108, 169, 170
falls, 81–82
family history, 20
fat, 98, 99, 101, 165, 245
fear, 61, 78, 118, 162, 164; aggression, 140–41, 147–48, 169; phobias, 172
feeding, 10, 31, 37, 39, 96–108, 133, 142, 152; and allergies, 110, 157–61, 177–78, 245; balanced diet, 16, 97–99, 100, 165, 217, 236; canned food, 99, 236; costs, 10; dieting, 103–105; digestive disorders, 159–66; do's and don'ts, 100; dry food, 37, 64, 96, 97, 99; eating disorders, 97, 100–108, 164–66; elderly cats, 236, 237, 239–43, 245; kittens, 28, 39, 45, 64, 96–97, 99, 100, 105–106, 107; and life span, 233–34; nursing, 39, 45, 96, 105–106, 107; nutritional needs, 98–99; and parasites, 183–85; preferences, 96–98; pregnant queens, 216–17; on prey, 70–73, 96; as reward, 116, 117, 119; sweets, 39, 64, 96; and travel, 87–88, 89; water, 99–100, 111
feet, biting, 168, 170
feline immunodeficiency virus (FIV), 161, 182, 187
feline infectious peritonitis (FIP), 161, 182, 188, 189, 228
feline leukemia (FeLV), 132, 161, 163, 182, 183, 185–87, 189, 190–91, 214, 228, 244
feline urinary disease, 54

feline urological syndrome, 110–11, 130, 132
female cat(s), 21–22, 194–97; birth weight, 222; and dominance, 76, 148, 204; eating habits, 97; genitals, 195, 196, 200–202, 206–207, 214–16; heat and breeding behaviors, 22, 36–37, 38, 39, 130, 194–97, 199, 200–202, 203, 205–208, 214–16; and hunting, 72, 82; identification of, 31–32; life span, 232–33; neutered, 9, 10, 16, 21–22, 38, 74–75, 131, 134, 197, 206–208, 244; sexuality of, 194–97, 200–202, 205–206, 214–16; sexual problems, 205–206; sisterhood, 73; spraying, 134–39, 195–96. *See also* mother cat(s)
feral cats, 72–73, 98, 176
fetal development, 214–18
fetching, 120–21
fever, 164, 176, 188
fighting. *See* aggression
fish, 70, 72
fleas, 62, 88, 160
Flehmen, 39, 196, 201
free-roaming cats, 72–74, 82–83, 93, 199
friends, leaving cat with, 92
funerals, 261–63

Gaddis, Vincent and Margaret, *The Strange World of Animals and Pets,* 40
games, 122–23
gas, 239–40
gender, 8, 9, 21–22; and dominance, 75–77; of kittens, 31–32; and life span, 232–33; selection, 21–22; and sex, 195–210; and social scene, 73–77. *See also* female cat(s); male cat(s)
genetics, 20–21, 204, 233; predisposition to genetic disorders, 212–14
genital(s): female, 195, 196, 200–202, 206–207, 214–16; grooming, 109; male, 198–205, 207; opening, 31–32
grieving process, 12–13, 252, 253–66
grooming, 8, 9, 10, 11, 45, 47–48, 57–65, 176; administering medications, 65–67; bathing, 63–64; behavior, 108–109, 167–72; claws, 57–61; displacement, 109, 167–72; elderly cat, 58, 132, 237; eyes and ears, 65; hair, 61–64; problems, 109; teeth and gums, 64–65
gums, 64–65
Guyot, Gary, 26

hair balls, 62, 161, 162, 163, 239, 240
hair coat, 21, 54, 61–64; brushing and combing, 61–63; of elderly cat, 236, 237, 238, 239, 240, 242; growth cycle, 61; length of, 9; licking, 108–109, 167–72; pulling, 168, 177, 213; shampooing, 63–64; shedding, 61
hair rings, 203–204
handling kittens, 24, 28–29
harness, 88, 126
health insurance, 11, 93–94
hearing, 21, 34, 41, 238
heart disease, 158, 214, 236, 239
heart worm, 158
heat, 22, 36–37, 38, 39, 130, 194–97, 199, 200–208, 214–16; cycles, 194–97, 207, 214, 215, 239
heat stroke, 88
hiding places, 78–79
high-rise syndrome, 81–82
Hill's Pet Products, 165
Himalayans, 21, 157–58, 170
hissing, 49, 50, 146
Holzworth, Jean, *Diseases of the Cat,* 171, 172
home range, 77–83
hormones, 135–36, 147; imbalance, 101, 102; of older cat, 181, 237; sex, 194–209, 214–16, 224; thyroid, 181–82
hotels, cats in, 86–87
Houpt, Katherine, 59
house, cat, 78–79
houseplants, eating, 106–107, 180
housesoiling, 12, 55, 131–39, 151, 237
human/feline physiological age comparison, 235–36
hunting, 25–26, 35, 42, 70–72, 77, 82, 143–44, 156; bird, 70, 71, 72, 96; discouraging, 72; encouraging, 71–72; rodent, 6, 7, 11, 25–26, 70–72
hydrocephalus, 213
hyperthyroidism, 181, 242
hypothyroidism, 182, 237, 242

ID collar, 83, 84
illness(es), 12, 36, 37, 54, 156; administering medications, 65–67; and behavioral problems, 130–33, 175–91; chronic, 16, 97, 220; digestive disorders, 159–66, 213, 214, 239–40; in elderly cat, 238–46; fatal, 250, 252; and housesoiling, 132–33; stress-related, 156–72; urinary, 54, 55, 110–11, 132, 133, 205, 242. *See also* disease(s); *specific illnesses*

Index

imitation, 42
inbreeding, 212, 235
independence, 7
indoor cat(s), 78, 81–82, 134, 234; declawed, 59–61; and dominance, 74–77; house for, 78–79; housesoiling, 12, 55, 131–36; intolerance to strange cats, 74–77; pedicure for, 57–58; male, 204–205
inflammatory diseases, 182
injuries, 46, 78, 164; and aggression between males, 146–47; and CPR, 67–68; fatal, 252, 257, 263–64; from falls, 81–82; trachea, 158
insectide poisoning, 178
insulin, 241–42
insurance, pet, 11, 93–94
intelligence, 41–44
intolerance of resident cats to strange cats, 74–77
isolated kittens, 24–25

jumping on counters, 151–52
jumping through hoop, 121
Jung, Carl, 8

Karsh, Eileen, 12
kennel, 89, 91; boarding, 92–93
kidney disorders, 161, 163, 182, 214, 235, 236, 239, 242, 246
kitten(s), 22, 62, 125, 195, 197; aggressive, 140, 144–45; aggressive parents of, 225–27; attachment period, 27; bathing, 63–64; behavior learned from mother, 25–26, 28, 42, 52–53, 70, 108–109, 222; birth of, 217–21; claws of, 58, 59, 149–50; death of, 226–28; deprived, 23; destructive, 149–50; diarrhea and vomiting, 160, 164; drinking, 99; excessive sucking, 105–106; exposure to humans, 28–29; feeding, 28, 39, 45, 64, 96–97, 99, 100, 105–106, 107, 221–23, 228–29; feline leukemia, 185; fetal development, 214–18; first weeks, 24–25, 221–23; handling and other stimulation, 24, 28–29; ill, 226; isolated, 24–25; and litter box, 52–54, 222; litter number and size, 214, 215, 218, 221, 226; locomotor development in, 41; mother's recognition of, 224–25; nursing, 39, 45, 96, 105–106, 107, 217, 218, 221–23, 228–29; orphan, 52–53, 105, 147, 206, 223; play and hunting, 25–27; potty training, 52–53, 54; selection of, 20–23; self-grooming, 108; sensory development in, 34–41; sexual development in, 31–32, 195–200, 203, 204, 222; and sisterhood, 73; socialization of, 26–32; and stress, 156, 160, 161; swimming, 125–26; teeth and gums, 64; traveling, 86–87; tree-climbing, 80–81; two, 22–23; vaccinations, 161; vision of, 34–35, 40, 221; and warmth, 37, 223; weaning, 24, 25, 27, 96, 105, 197, 217, 222–23
KMR (Pet-Ag), 160, 228
Kübler-Ross, Elisabeth, 254

labor and birth, 217–21
lactation, 96, 156, 217, 218, 221–23, 228–29
language, 45–50; body, 38, 46–49, 140, 146; purring, 45–46, 47, 49, 50; vocal communication, 46
laxatives, 67
lead poisoning, 130, 178–79
learning, 41–45; from mother cat, 25–26, 28, 42, 52–53, 70, 108–109, 222; names, 44–45; observation and imitation, 42; problem solving, 43–44; trial and error, 43. *See also* training
leash, 88, 126; laws, 57
license, 10
licking, 50, 108–109; compulsive, 167–72, 178–79, 215; kittens, 221–23
life span, 232–36
limping, 173
line breeding, 212
liquid medications, 67
litter box, 9, 10, 16, 31, 36, 37, 52–57, 66, 176; burying feces, 52; cleaning, 53–55, 184; commode as, 55–56; and kittens, 52–53, 54, 222; outside, 56–57; problems, 37, 54, 56–57, 117–18, 130, 131–39, 151, 159, 237, 241; suggestions, 54–55; symptoms of litter aversion, 136–37; and traveling, 84, 85, 87; types of, 53–55
liver disorders, 161, 163, 182, 236, 239, 242
living spaces, changes in, 130, 146
locomotor development, in kittens, 41
longevity, feline, 232–36
long-haired cats, 21, 54, 132–33; grooming, 62–63
love/hate relationship, between people and cats, 6–8
lymphosarcoma, 163

magnesium, 111
male cat(s), 21–22, 138; aggression between, 146–47, 199, 202, 204, 206, 208; birth weight, 222; brotherhood, 73–74; dominance in, 75–76, 138–39, 146–49, 202–205, 226–27; genitals, 198–205, 207; hunting, 72, 82; identification of, 31–32; life span, 232–33; neutered, 9, 10, 16, 21–22, 38, 72, 82, 131, 134, 144, 145, 147, 198, 203, 204, 206–208, 210; as parent, 226–27; Plugged Tomcat Syndrome, 110–11; roaming, 72–74, 82–83, 93, 199, 206; sexuality, 198–205, 214–15, 226; sexual problems, 203–205, 210; spraying, 21, 38, 77, 134–39, 199; tomcat behavior, 72–76, 82, 93, 140, 144–47, 187, 197, 198–205, 206, 208, 226–27; weaned early, 25
Manx, 213, 214
marijuana, 178
matching feline and human personality types, 13–14
meat, 98–99
medication(s), 92, 93; administering, 65–67, 116; anticonvulsant, 183; cancer, 244–45; poisoning, 178–80; for spraying, 135–36; thyroid, 181–82; tranquilizers, 87, 135, 136, 142, 151, 162, 166, 167–68, 259
Medipet, 94
memory, 43–44
mental ailments, stress-related, 166–73
meowing, 46, 50
Middle Ages, cats in the, 7
milk, 100, 110, 160; intolerance, 160, 245; lactation, 96, 156, 217, 218, 221–23
misbehavior. *See* behavioral problems
morning-after shot, 208
mother cat(s), 37, 52, 62, 131; age of, 214; aggressive, 140, 225–26; behavior learned from, 25–26, 28, 42, 70, 108–109, 222; care of prospective mother, 216–17; and deprived kittens, 23, 25; hunting skills, 70, 71; labor and birth, 217–21; and litter box, 52–53, 222; litter number and size, 214, 215, 218, 221, 226; nursing, 39, 45, 96, 105–106, 107, 217, 218, 221–23, 228–29; pregnancy, 12, 156, 194, 195, 201, 207, 214–21, 225, 227, 228–29; recognition of her young, 224–25; reproductive system, 194–97, 201, 214–21, 227–28; retrieving behavior, 223; and weaning, 24, 25, 27, 96, 105, 197, 217, 222–23
motion sickness, 162
mouth: disorders, 213, 243, 245; open and panting, 49
mummified cats, 7, 260

naming, 44–45, 118–19
napping, 34, 109–10
National Animal Poison Control Center (NAPCC), 179, 180
nausea, 162
negative reinforcement, 114–15
nervousness, 13, 47, 130, 163, 167; and obsessive-compulsive behaviors, 166–72
nervous system, 35, 98, 179, 238
neutering, 9, 10, 16, 21–22, 38, 72, 74–75, 82, 93, 101, 131, 134, 144, 147, 197, 198, 203, 206–208, 234; advantages of, 208; female, 9, 10, 16, 21–22, 38, 74–75, 131, 134, 197, 206–208, 244; male, 9, 10, 16, 21–22, 38, 72, 82, 131, 134, 144, 145, 147, 198, 203–208, 210; and obesity, 101, 102; procedure, 206–207; and spraying, 134, 135, 206, 208
Neville, Peter, *Do Cats Need Shrinks?*, 139
night vision, 35–36, 97
nipple preferences, 221
nose, 37–39
number of cats, 8; and dominance, 75–77; and feline leukemia, 186–87; and housesoiling, 132, 134; two, 22–23
nursing, 39, 45, 96, 105–106, 107, 217, 218, 221–23, 228–29
nutritional needs, 98–99
nymphomania, 206

obesity, 97, 100–105, 161, 182, 220, 233, 235, 236, 239, 240–42
objectionable mounting, 203, 208
observation, 42
obsessive-compulsive disorders, 166–72, 178
oral medications, 65–67, 116, 166
outdoor cat(s), 70–83, 108, 134; declawed, 60; and home range, 77–83; hunting, 70–72; intolerance of resident cats to, 74–77; rank and territory, 74–77, 145–46, 199; roaming, 72–74, 82–83, 93, 199, 206; social scene of, 72–77; tree climbing, 80–81
ovaries, 195, 201, 206, 207, 208, 214–15
ovulation, 201, 205, 215

pancreas, 241, 242
panleukopenia, 161, 163, 189, 228
parasites, 88, 160–61, 182–85, 187, 214, 228, 234
parenting, 211–29; female, 221–26; labor and birth, 217–21; male, 226–27; mating, 193–210; pregnancy, 214–21, 225, 227, 228–29; reproductive failure, 227–28. *See also* mother cat(s)
paste medications, 67
paws, 109; kneading, 48. *See also* claws
pedicure, 57–58, 149, 237
pedigree, 212
penis, 32, 198, 201–205
people-cat relationships, 5–18; bonding in, 14–18, 248–67; choice of cat in, 8–11, 20–23; and death of cat, 247–67; love/hate in, 6–8; problems in, 11–14
Persians, 21, 42, 130, 206, 212, 213
personality, 9, 13–14, 20, 49–50, 102; and breed, 20–21, 212; matching feline and human types, 13–14
pet-loss support groups, 257–58
Pets Are Inn, 92
pet-sitting services, 91–92
phobias, 172
physical characteristics, 9, 21; and sex, 199–200
pills, administering, 65–67
placentas, 219
plant eating, 106–107, 180
play, 25–27, 236; rough, 144–45
play aggression, 143–44, 147
playing dead, 122
Plugged Tomcat Syndrome, 110–11
poisoning, 106–107, 130, 161, 164, 178–80
polydactylism, 212
polygamy, 199–200
positive reinforcement, 115–17, 150
potassium deficiency, 242
pregnancy, 12, 156, 194, 195, 201, 207, 214–21, 225, 227, 228–29; detection of, 216; false, 205; gestation period, 215–16, 220; labor and birth, 217–21; prevention of, 206–209; reproductive failure, 227–28
pregnant women, and toxoplasmosis, 183–85
Prescription Diet, 165, 245
prey, 25, 35, 36, 42, 52, 70–72, 96
problem relationships, 11–14
problem solving, 43–44
protein, 98, 99, 165, 217
Pryor, Karen, 114
pseudopregnancy, 205
psychic sense, 39–40
psychogenic skin disease, 170
puberty, 196–99, 214, 235, 244
punishment, 7–8, 114, 117–18, 135, 150
purring, 45–46, 47, 49, 50, 221
pyloric dysfunction, 163, 213

quarantines, 189–90
queening, 217–21

rabbits, 70, 73
rabies, 176–77, 189, 190
rank and territory, 74–77, 145–46
redirected aggression, 142–43, 147
redirected love, 14
registered cats, 44, 212
regurgitation, 162–63
reinforcement, 115–17, 150
replacement pets, 266
reproductive system, 194–97, 201, 214–21; disorders, 185, 188, 227–28, 239. *See also* sex
resident cats, and intolerance to strange cats, 74–77
respiration, in elderly cats, 238, 244
retching, 162
retrieving behavior, 223
rewards, 116–17, 119
Rex, 20, 21
rhinotracheitis, 158, 189, 228
ringing doorbell, 121–22
roaming cats, 72–74, 82–83, 93, 199, 206
rodent bait poisoning, 179
rodent hunting, 6, 7, 11, 25–26, 36, 52, 70–72
rolling skin disease, 169, 170–72
Russian Blue, 21

Savishinsky, Joel S., 249
scent, 30, 37–39, 108; greeting, 38; marking, 38, 77, 131, 134
scratching, 16, 149; as behavioral problem, 130, 140, 144, 147, 149–51, 168–72, 178–79; territorial, 77, 149
scratching post, 58–59, 149–50
scrotum, 31, 198
seizures, 182–83
selection of cat, 8–11, 20–23; age, 22; deprived kitten, 23; gender, 21–22; genetics and breeding, 20–21; physical characteristics, 21; two cats, 21–23
senses, 34–41; development in kittens, 34–35, 40–41; in elderly cat, 238; hearing, 34, 41, 238; psychic, 39–40;

smell, 37–39, 41, 77, 238; taste, 39, 41, 238; touch, 36–37, 41; vision, 34–36, 40, 77, 238
separation anxiety, 156; destruction, 151
sex, 6, 25, 45, 73, 75, 76, 77, 82, 130, 134, 193–210; and aggression between males, 146, 198–200, 202–208; behavioral problems, 130–31, 200, 202, 203–206; and breeds, 212–14; female, 194–97, 200–202, 205–206, 214–16; intercourse, 199, 200–203, 214–15; male, 198–205, 214–15, 226; mating calls, 45, 49, 134, 195–96, 200–201; polygamous, 199–200
shaking hands, 122
shampoo, 16, 63–64, 209
shedding, 61
short-haired cats, 214; grooming, 62
showing, 11, 59, 156
Siamese cats, 20, 21, 42, 49, 107, 130, 157, 167, 170, 206, 213–14, 235, 244
sibling(s), 25; litter size, 214, 215, 218, 221, 226; rivalry, 147
sisterhood, 73
sit (command), 119
sitters, cat, 91–92
size of cat, 8, 9
skin, 62, 63, 167; disorders, 170–72, 177–78, 213, 238, 243, 244
sleep, 109–10, 239; and fever, 176
smell, 37–39, 41, 77, 238
Smith, Scott S., 251
socialization, 9, 13–14, 26–32, 72, 130, 145, 156; attachment period, 26; exposure to animals, 30–31; exposure to humans, 28–29
social scene, 72–77; brotherhood, 73–74; dominance in, 75–77; feral cats, 72–73; intolerance of resident cats to strange cats, 74–77; rank and territory, 74–77; sisterhood, 73
Soft Paws, 60, 61
soul, 251–52
spaying. *See* neutering
sperm, eggs fertilized by, 14–15
spina bifida, 213
spaying, 21, 38, 77, 130, 131, 132, 134–39, 195–96, 199, 206; and neutering, 134–35, 206, 208
stairs, climbing, 124–25
stay-at-home cat, 90–92
stop (command), 119–20
strange cats, intolerance of resident cats to, 74–77
stress, 36, 91, 130, 155–73, 253, 265; and anorexia, 105, 164–66; and compulsive behaviors, 167–72; and coughing, 157–58; and diarrhea, 159–61, 164; effects of, 156–57; in elderly cat, 237; and grooming behavior, 109, 167–72; and kittens, 156, 160, 161; and mental ailments, 166–73; -related disorders, 157–72; and sex, 130–31, 202–206, 226; and vomiting, 161–164
stud tail, 209
sucking, excessive, 105–106
sunlight, 78–79
surgery, 164, 167, 173, 181, 205; neutering, 197, 206–207
sweat, 38, 77, 108
swimming, 125–26
symbols, cat, 6–7

tables, jumping on, 151–52
tail, 196, 200, 202; biting, 168, 170; language, 47, 49, 61, 140, 144; stud, 209
tapeworm, 160
taste, 39, 41, 238
taurine, 98, 227
teaching. *See* learning; training teeth, 64–65, 70, 108–109; of elderly cat, 236, 237, 239
territory, 74–77, 199; and aggression, 145–46; marking, 77, 131, 134, 138–39, 149, 199, 208
testicles, 31, 198, 199, 205, 207; retained, 204, 212
test litter, 54
testosterone, 198, 199, 205, 237
tethering, 126–27
third eyelid, 35
thyroid gland, 102; disease, 181–82, 237, 242
ticks, 88
toilet training, 31, 52–57. *See also* litter box
tongue, 50, 108–109
touch, 36–37, 41
toxoplasmosis, 182, 183–85, 228
toys, 26–27, 105, 106, 122–23
trachea, 158
training, 10, 113–27; collar and leash, 88; commands, 118–20; for compulsive behaviors, 169; etiquette, 52–68; exercise, 123–24; games, 122–23; methods, 114–15; and punishment, 114, 117–18, 135, 150;

training (*cont.*)
reinforcement, 115–17, 150; rewards, 116–17, 119; tricks, 120–22
train travel, 90
tranquilizers, 87, 135, 136, 142, 151, 162, 166, 167–68, 259
travel, 83–90; air, 85, 88–90; boat, 90; bus, 90; car, 27, 31, 83–88, 162; carrier for, 84–86, 88–89; and stay-at-home cat, 90–92; train, 90
tree climbing, 80–81
trial and error, learning by, 43
tricks, 27; teaching, 120–22
tumors, 243–45
two cats, 22–23; introduction to each other, 30–31

University of Pennsylvania, 252, 253
urethra, 110–11, 205
urinary disease, 54, 55, 110–11, 132, 133, 205, 242
urinary tests, 54
urination, 52–57, 99; bed wetting, 138–39, 157; on carpets, 12, 55, 133, 137; in commode, 55–56; in elderly cat, 237, 239, 241–42, 243; housesoiling, 12, 55, 131–39; and kittens, 52, 222; and litter box, 52–55; Plugged Tomcat Syndrome, 110–11; spraying, 21, 38, 77, 130, 131, 132, 134–39, 195–96, 199, 206, 208; as territorial marking, 38, 77, 131, 134, 138–39, 208. *See also* litter box
uterus, 195, 206, 207, 208, 215, 220

vaccinations, 9, 10, 46, 84, 93, 141, 158, 161, 176–77, 185–89, 214, 234, 244; schedule for, 189
vagina, 195, 201
Valium, 135, 136, 142, 166, 167–68
veterinarians, 10, 16, 27, 35, 54, 60, 65, 87, 89, 91, 93–94, 97, 104, 110–11, 114, 131, 132, 135–36, 140, 152–53, 160, 164, 165, 178, 179, 181, 186, 206–207, 246; and birth of kittens, 214, 216, 219–21; and death of cat, 254–60, 266; second opinions, 246
Veterinary Pet Insurance, 94
viral infections, 158, 161
viral rhinotracheitis, 158, 189, 228
vision, 34–36, 40, 77, 238; night, 35–36, 97
vitamins, 98, 217, 236; deficiency, 182, 183
vocal communication, 46
Voith, Victoria, 14
vomeronasal organ, 38–39
vomiting, 99, 106, 157, 178, 179, 244, 245; and pregnancy, 216, 218; stress-related, 161–64
vulva, 32, 195, 196, 200, 201, 202; and birthing, 218–21

Walters, Mark, *The Dance of Life: Courtship in the Animal Kingdom,* 199–200
water, 99–100, 111, 164–65; and travel, 87, 89
water bug, 218–19
weaning, 24, 25, 27, 96, 105, 197, 217, 222–23
weather changes, 170–72
weight, 9, 114, 135; and anorexia, 105, 164–66, 235, 245; birth, 221, 222; and obesity, 100–105, 235, 236, 239, 240–42
Wexler, Toby, 60
whiskers, 36, 61, 71, 97; language, 48
white cats, 21, 204, 213, 244
wool chewing, 107–108, 157, 169, 170
worms, 158, 160, 163, 214

x rays, 216

Printed in the United States
71468LV00003B/156

9 780865 345096